# Fuels and Petrochemicals Division 2020



Held at the 2020 AIChE Spring Meeting and 16th Global Congress on Process Safety

Online
17 - 21 August 2020

ISBN: 978-1-7138-2346-9

**Printed from e-media with permission by:**

Curran Associates, Inc.
57 Morehouse Lane
Red Hook, NY 12571

**Some format issues inherent in the e-media version may also appear in this print version.**



Printed with permission by Curran Associates, Inc. (2021)



AIChE
120 Wall Street, FL 23
New York, NY 10005-4020

Phone: (800) 242-4363
Fax: (203) 775-5177

www.aiche.org

**Additional copies of this publication are available from:**

Curran Associates, Inc.
57 Morehouse Lane
Red Hook, NY 12571 USA
Phone: 845-758-0400
Fax: 845-758-2633
Email: curran@proceedings.com
Web: www.proceedings.com

# TABLE OF CONTENTS

# 591161 2020 AIChE Spring Meeting & 16th GCPS AGILE Award Keynote Address - Beyond the Horizon: The Factors and Forces Affecting the Global Petrochemical Industry

**Abstract:**

Over the past 10 years, LyondellBasell and the global petrochemical industry has undergone a dramatic transformation. In this era of unprecedented disruption, it is likely that the next the decade will present both challenges and opportunities for the petrochemical sector. LyondellBasell CEO Bob Patel will summarize the company's progress to date and provide a forward-looking perspective on key opportunities and risks faced by the industry in 2020 and beyond.

# 592550 Digital Transformation in Chemical Industry

**Abstract:**

Manufacturing is the original denizen of one of the cornerstones of Industry 4.0 - a digital transformation journey to unlock the full potential of manufacturing data. This talk will highlight the critical human element to deliver sustainable and scalable digital transformation, discuss common challenges using Artificial Intelligence and Advanced Analytics as an example of digital transformation, focusing on corporate culture change. A proven best practice will be shared to illustrate real-world outcome in digital transformation.

# 591157 Securing Enterprises via Cybersecurity Understanding, Resiliency, and Education

**Abstract:**

**Securing Enterprises via Cybersecurity Understanding, Resiliency, and Education**

Industry 4.0 initiatives are enabled by leveraging the deployment of information technologies into the operational technology environment. While a logical approach in defining the future of smart manufacturing, it remains fret with challenges and risks that must be mitigated assure enterprise resilience to cyber and cyber-physical attacks. As economies recalibrate in the aftermath of the current public health emergency, definitions of what constitutes "critical infrastructure" are broadening while, at the same time, attack surfaces and the threat landscape are expanding. This means companies must place a greater emphasis on taking the steps and making the investments necessary to overcome the challenges and navigate these challenges and risks to maximize their resilience and secure and protect their enterprises. This presentation will explore the link between cybersecurity and process safety in the context of a safety culture. It will also identify cybersecurity challenges in industry 4.0 initiatives, the integration of information technologies (IT) and operational technologies (IT), the delivery of training, and the need for credentialing and ongoing professional development.

# 590270 Mo Doped Novel Solid Acid Catalysts: A Comparative Study for Methane Aromatization

**Abstract:**

**Mo doped novel solid acid catalysts: A comparative study for methane aromatization**

Ashraf Abedin[1], Swarom Kanitkar[1,2], Srikar Bhattar[1], James J. Spivey[1]*

[1] *Cain Department of Chemical Engineering, Louisiana State University, Baton Rouge, LA, USA*

[2] *National Energy Technology Laboratory, U.S. Dept. of Energy, Morgantown, WV 26507*

*jjspivey@lsu.edu (corresponding author)

The production of natural gas in the United States is now at its peak, due to novel technologies such as horizontal drilling and hydraulic fracking. The primary end uses of natural gas include power generation and flaring. Methane is the most abundant constituent of natural gas, and has the potential to produce high value chemicals. This includes valuable products such as benzene, which is currently obtained by crude oil processing through catalytic reforming. Typical methane conversion processes consist of indirect reactions that include co-reactants, which makes these processes less energy efficient. Direct conversion of methane does not involve an intermediate step.

One such direct pathways to activate methane is methane aromatization (MDHA), which converts methane directly to benzene and hydrogen.

$$6CH_4 = 4C_6H_6 + 9H_2; (\Delta G_{298K} = +433 \text{ kJ/mol}, \Delta H_{298K} = +530 \text{ kJ/mol}) \qquad \ldots (1)$$

Very high temperatures ($700^0$C-$1000^0$C) are needed to drive the equilibrium towards the formation of benzene and hydrogen. But at these high temperatures, thermodynamics are selective towards formation of coke. This makes MDHA as one of the most challenging problems in $C_1$ catalytic chemistry. Mo oxide supported on ZSM-5/MCM-22 has been well-studied in recent years for MDHA. It is generally thought that Mo carbides are responsible for activating methane by forming $CH_x$ species, which are dimerized into $C_2H_y$ species and then oligomerized on the acidic zeolite sites on ZSM-5/MCM-22, leading to aromatics.

Although homogeneous acid catalyzed-reactions are typically active and selective, the separation of the products from the catalysts is expensive. Solid acids avoid the separation step, and can be used at an elevated temperature. In this work, we have tested different solid acids as supports for Mo oxide to compare the activity for methane conversion with in-house prepared ZSM-5 (Si:Al=50:1). Here, sulfated zirconia (SZ) and sulfated hafnia (SH) are prepared as novel acid supports for Mo oxide. After loading Mo, these catalysts were characterized using Raman, pyridine DRIFTS and ammonia TPD before MDHA reactions were run.

Three catalysts were tested: Mo/SZ, Mo/SH, and Mo/ZSM-5. All the three catalysts showed comparable catalytic activity for MDHA runs at $650^0$C. Mo/SZ showed higher selectivity to benzene (Figure 1), which was comparable to that of Mo/SH. In comparison to these two solid acids, Mo/ZSM-5 was not highly selective towards benzene at $650^0$C. This may be due to a higher number of stable acid sites for SZ and SH based Mo catalysts in comparison to that of ZSM-5 (Table 1). Deactivation was observed for all the three catalysts with time. This can be primarily attributed to carbon formation at catalytic surfaces, which was later confirmed with subsequent TPO analysis.

Table 1: Ammonia-TPD results for the three catalysts

| **Catalyst** | **Total micromoles of $NH_3$ desorbed/ consumed per gram catalyst** |
|---|---|
| Mo/SZ | 1560.0 |
| Mo/SH | 1280.7 |
| Mo/ZSM-5 | 1143.4 |

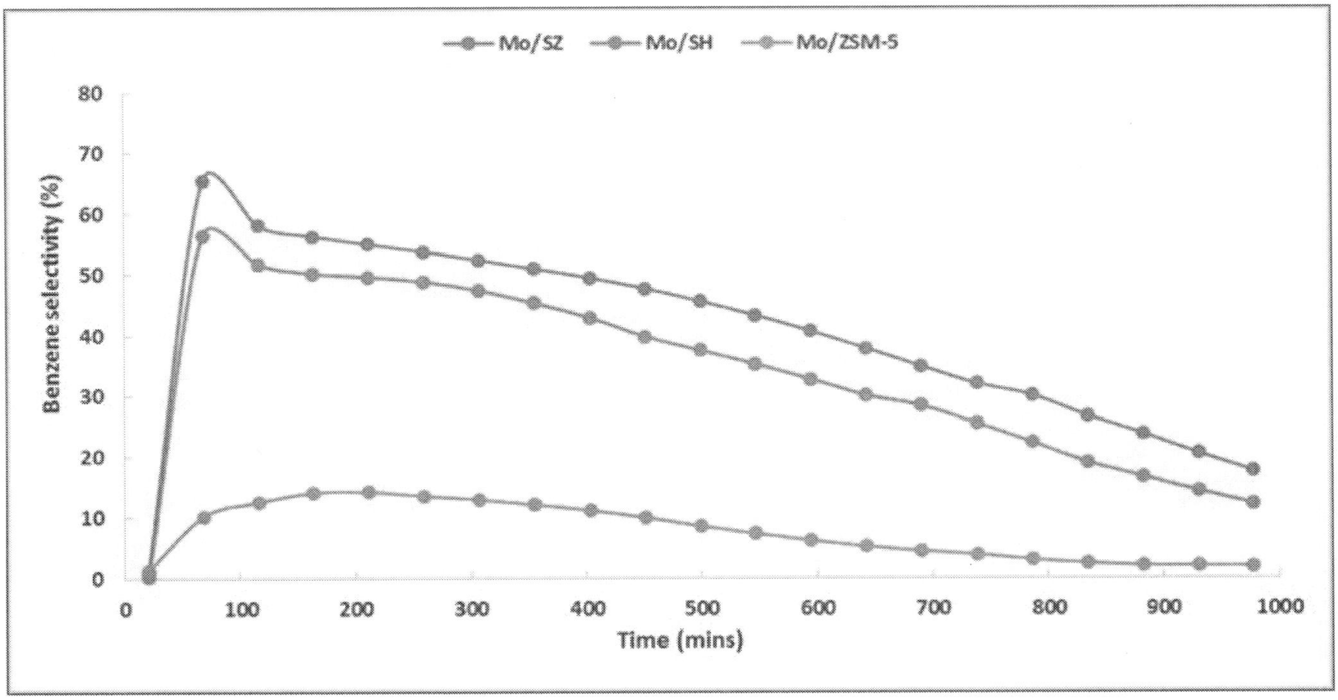


Figure 1: Benzene selectivity for methane aromatization for Mo doped different solid acid catalysts. ($650^0$C, 10 SCCM methane, 1 atm, 500 mg)

# 589836 Hazard Identification and Risk Analysis for Retrofitting a Biorefinery

**Abstract:**

One important aspect of future green technologies includes biorefining of renewable feedstocks to make energy, fuels, chemicals, and other useful products. One of the major challenges associated with the development of biorefineries is the large capital cost and risk associated with building a production plant for a technology that has not previously been proven on a commercial scale. Retrofitting existing process plants can serve to alleviate these risks. Specifically, the modification of existing first generation biofuel processes, or integrating new technologies into pulp and paper mills can be attractive options for hedging potential risk and lowering cost. However, in the implementation of the retrofit, process safety issues related to feedstock handling, uncharacterized waste streams, combustible dust, and design limitations from existing equipment can become a serious issue. Herein, the process safety issues associated with retrofitting an existing plant for biorefining are addressed. Examples of potential hazards associated with biorefining processes with new feedstocks, chemistries, and processing units are discussed as well as recommendations for process hazard assessment methodologies related to retrofitting existing process plants for biorefining.

# 589838 Techno-Economic Analysis of Direct Methane Aromatization for Natural Gas Upgrading with Mini-Plant Construction for Remote Deployment

**Abstract:**

Inexpensive sources of natural gas have led to a boom in development of technologies taking advantage of C1 chemistry; however, a significant number of potential natural gas sources are stranded and lack the infrastructure for economically-feasible transportation. The resulting gas which is normally flared can, in theory, be recovered at the well-site, purified, and then processed to a product which is easier to transport. Gas-to-liquid (GTL) processes exist which can convert the feed to liquid products which can be more easily transported, although there are process constraints due to the need for deployment in remote locations, and several are being commercialized.[1] One such process, direct methane aromatization (DMA), produces significant amounts of benzene, a valuable commodity chemical, and hydrogen. The benzene can be purified on-site and then transported by truck; the hydrogen can be dealt with on-site as well by either transforming it to another transportable chemical, such as ammonia or cyclohexane, or using it to provide heating or electricity.

Previous economic studies of DMA have been reported which either invoke unrealistic assumptions in the process model,[2, 3] do not consider use of the gaseous hydrogen,[2, 3] or do not consider the limitations imposed on the process due to remote deployment.[4] With these issues in mind, a DMA mini-plant processing 1.0 MMSCFD of natural gas was designed with currently available technology to provide a platform for the investigation of process intensified alternatives. The reactor section of the plant uses a pre-carburized DMA catalyst developed at Texas Tech.[5] The unit was investigated for economic feasibility and sustainability. In particular, the analysis identifies four particular areas which can benefit from intensification: aromatics yield (i.e. one-pass conversion), coke selectivity, hydrogen recovery, and separation and purification of the aromatic products. Although intensified reactor technology for the DMA process is already being investigated for improving the yield with[6, 7] and without integrated hydrogen recovery[8, 9], the process model framework should provide opportunities for investigating additional intensification technologies, particularly in separations.

1. H. Fleisch. [White Paper], World Bank-Global Gas Flaring Reduction Partnership, 2015.
2. Pérez-Uresti, S.I. *et al.*, *Processes*, 2017, 5, 33.
3. Camilo Corredor, E., P. Chitta, and M. D. Deo. *Fuel Pro. Tech.*, 2019, 183, 55 – 61.
4. Huang, K. *et al.*, *Joule*, 2018, 2, 349 – 365.
5. Rahman, M., A. Sridhar, and S.J. Khatib. *Cat. A Gen.*, 2018, 558, 67 – 80.
6. Kee, B. *et al.*, *Eng. Chem. Res.*, 2017, 56, 3551 – 3559.
7. Morejundo, S.H. *et al.*, *Science*, 2016, 353(6299), 563 – 566.
8. Brady, C., B. Murphy, and B. Xu. *ACS Catal.*, 2017, 7(6), 3924 – 3928.
9. Cao, Z. *et al.*, *Chem. Int. Ed.*, 2013, 52, 13794 – 13797.

# 589968 Combined Xpeem and Stxm Study of Coke Formation on High Temperature Al-Containing Alloy during Steam Cracking of Ethane

**Abstract:**

## Combined XPEEM and STXM study of coke formation on high temperature Al-containing alloy during steam cracking of ethane

Manjunath Patil[1], Jia Zhang[1], Brecht Laforce[2], Rachid Belkhou[3], Sufal Swaraj[3], Hilde Poelman[1], Vladimir Galvita[1], Laszlo Vincze[2], Guy B. Marin[1], Marie-Françoise Reyniers[1], Kevin M. Van Geem[1]

[1] *Laboratory for Chemical Technology, Ghent University, Technologiepark 125, 9052 Gent, Belgium.*

[2] *Department of Chemistry, Ghent University, Krijgslaan 281-S12, 9000 Gent, Belgium.*

[3] *Synchrotron Soleil, L'Orme des Merisiers, 91190 Saint-Aubin, France.*

*Keywords: ethane steam cracking, high-temperature alloys, coke formation, scanning transmission X-ray microscopy, X-ray photoemission electron microscopy*

Steam cracking of hydrocarbons is still one of the most vital chemical processes for the production of many key building blocks of the chemical industry. Almost 200 million tons of ethylene are produced globally per year, underlining the process significance [1]. During the steam cracking process, carbonaceous deposits develop on the inner wall of the tubular reactors due to high carbon activity, a process known as carburization. This reduces the cross-sectional area of the reactor, leading to pressure drop, and obstructs the heat transfer from the furnace to the reactor. With suitable pretreatment/oxidation of the material, the lifetime of the coil can be increased by creating an effective oxide barrier and thereby achieving reduced coking rates [2].

Limited information is available on these carbonaceous deposits on the reactor wall and their composition from the surface of the alloy to the bulk. Characterizing these materials requires adequate sensitivity and detection limits for the light carbon to be separated from the heavier metal elements. In view of this goal, soft X-ray techniques giving access to the light C element, are ideally suited to solve these scientific questions. Two such complimentary techniques, Scanning Transmission X-ray Microscope (STXM) and X-ray PhotoEmission Electron Microscope (XPEEM), were used at the HERMES beamline of the SOLEIL synchrotron in Saint-Aubin, France, to detail the alloy carburization process.

The coked materials were produced in the thermogravimetric unit, which allowed evaluating on-line the coking behavior of a fresh vs. pretreated coupon from Al-containing high-temperature reactor materials. The steam cracking experiments consisted of five cracking cycles with four hours ethane cracking at 950°C, followed by ramping-up to 1050°C for two hours per cycle, the latter resembling the end-of-run industrial conditions.

To understand the effect of pretreatment on the rate of coke formation of these Al-containing alloys (coupons) during steam cracking of hydrocarbons, off-line STXM, XPEEM and SEM-EDX measurements were performed on the coked samples. The objective of these experiments was to study the elemental distribution of carbon on/in the alloy matrix, as well as the elemental speciation for carbon and its main alloying constituents (Fe, Ni, Cr, Al) with nanoscopic spatial resolution (~40nm). For STXM, samples need to have a thickness of less than 100 nm with a depth of 30 μm, while tracking orientation. Three different techniques were applied to make such samples out of the coupon: FIB-SEM, microtome slicing and electropolishing. While the latter only leads to bulk coupon samples, FIB-SEM could deliver wedges, meeting all STXM requirements.

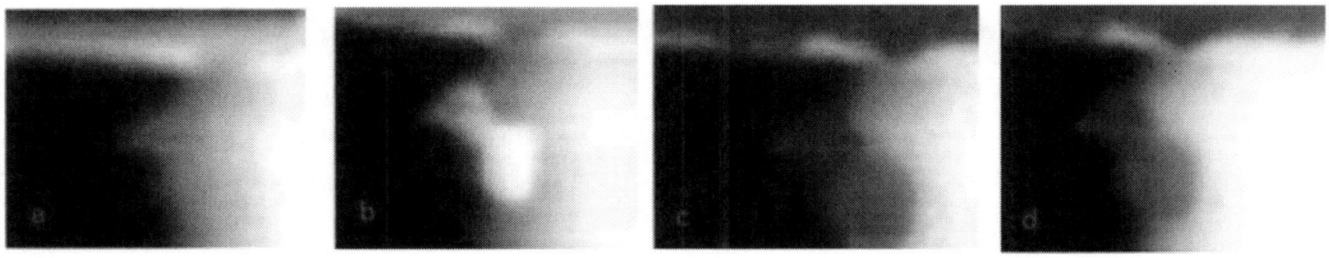


Figure 1 Images based on averaged x-ray absorption signals from the stack of the bottom of the pretreated steam cracked Al containing coupon for a: C K edge (~285 eV) , b: Cr $L_3$ edge (~575 eV), c: Ni $L_3$ edge (~853 eV) and d: Fe $L_3$ edge (~707 eV).

From the FIB-SEM samples, X-Y-I-E stacks (XY position, Intensity, Energy) were acquired for 4 elements, C K, Cr $L_3$, Fe $L_3$ and Ni $L_3$ edge, allowing to come to element mapping in the thin zone of the wedge. Concentration maps were obtained by averaging a set of images above and below an X-ray absorption edge, providing an improved signal to noise ratio. Figure 1 shows the result for C, Cr, Ni and Fe. As these images were made based on the X-ray absorption signal slightly above the element's edge, brighter intensity reflects more of this element is present. As such, a particle with higher Cr content is clearly visible at this position.

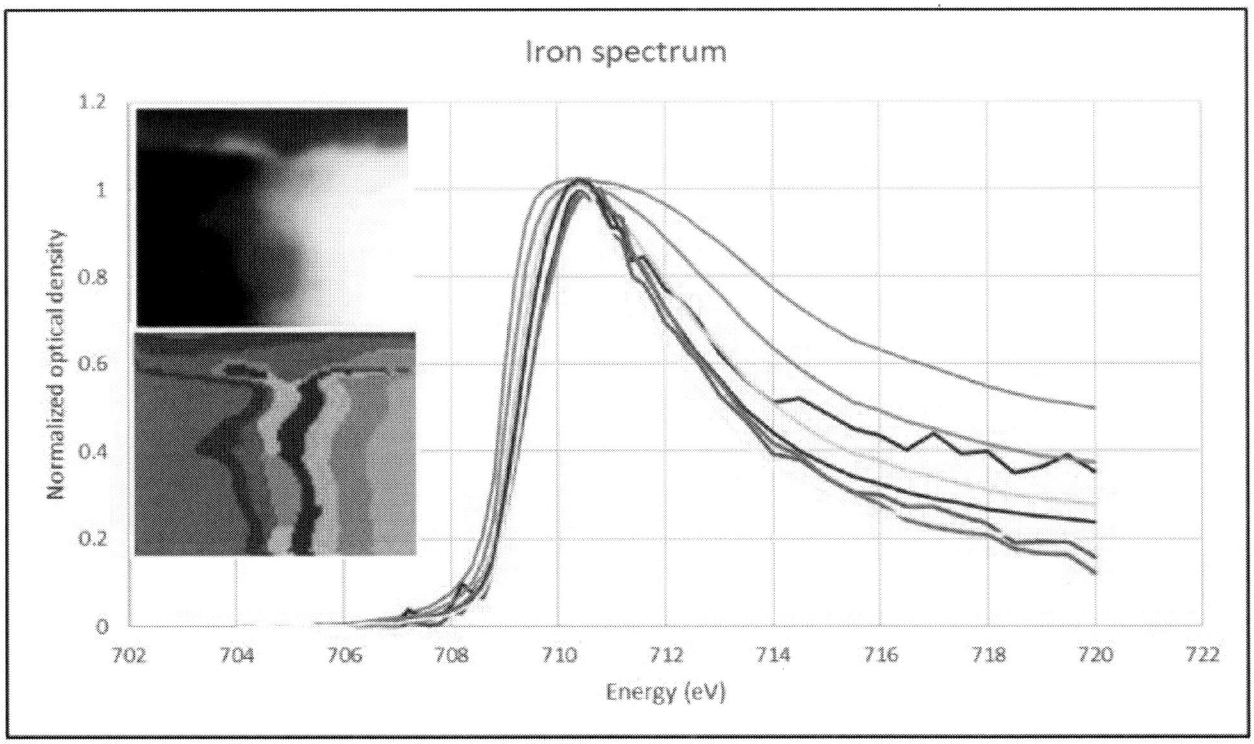


Figure 2 Normalized Fe $L_3$ optical density graphs vs. x-ray energy produced from STXM on the pretreated steam cracked Al containing coupon. Inset: STXM image of the coupon wedge near the bottom and concommittant PCA

cluster analysis for the Fe $L_3$ edge using MANTiS software. The spectrum color is equal to the color of the cluster.

Figure 2 shows the normalized Fe $L_3$ edge spectra obtained from STXM analysis of Figure 1d (inset) for a pretreated, steam cracked Al containing coupon. The stacked image 2.5 $\mu m^2$ with a step size of 25 nm was processed using PCA-based cluster analysis to identify clusters or zones with significantly different Fe $L_3$ signals. Each signal pertains to a cluster or zone of the image and will differ by a combination of varying sample thickness and chemical speciation of the element in that zone. These measurements are analyzed from the surface to the bulk (up to ~5μm) of the coupon.

In contrast to the transmission experiments of STXM, XPEEM looks at the emitted electron yield, hence with high surface sensitivity (probing depth: few nm). For the steam cracked coupon with and without pretreatment, such surface analysis was performed on 4 or 5 different positions along the coupon surface, to assess the difference in coking behavior of untreated vs. pretreated coupons. Figure 3 shows the curves obtained at the C K edge at 4 or 5 positions on both samples. Similar curves were recorded for the Ni $L_3$, Fe $L_3$, Cr $L_3$ and O K edge and will be analyzed in parallel to correlate the results.

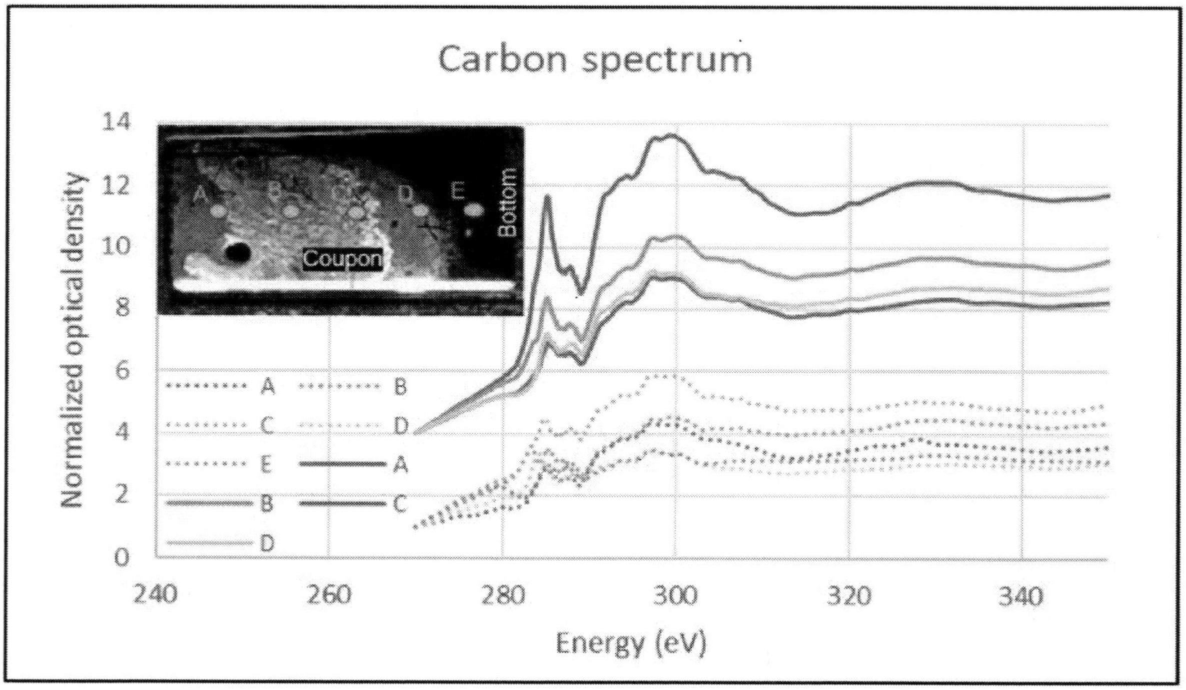


Figure 3: optical density curves acquired at the C K edge from the surface of steam cracked (dotted lines) and pretreated+steam cracked (solid lines) Al containing coupons, at different surface positions. Inset: SEM image of the non-pretreated steam cracked coupon with indication of the acquisition positions.

The results based on coking rates and SEM with EDX indicates that with the applied pretreatment conditions, the internal oxidation of alumina can be avoided and therefore coking rates are reduced by a factor 4 compared to non-pretreated coupons.

**References**


1. ExxonMobil, *The Outlook for Energy: A view to 2040.* 2012.

2. Sarris, S.A., et al., *Optimization of the in-situ pretreatment of high temperature Ni-Cr Alloys for Ethane steam cracking.* Ind. Eng. Chem. Res., 2017.

# 589775 Data Infrastructure As Foundation to Operational Excellence and Digital Transformation

**Abstract:**

Is technology a solution looking for a problem? A successful digital transformation builds on the foundations of Operational Excellence. Innovations in the technology we use further these initiatives. The building blocks of these improvements are rooted in the data we collect, compile, and contextualize into operational intelligence. Successful implementations identify desired outcomes and build the data and technical infrastructures to achieve them.

A structured Operational Excellence program begins with clear definitions for:

- What's important?
- Who's responsible for action, who's accountable for success?
- How is success measured?
- How are excursions identified, monitored, and corrected?
- What communications are necessary to ensure improvement?
- What skills and tools are necessary to continuously improve?

Compelling events for transformation may include regulatory compliance, hazard recognition, sustainability, cost controls, quality improvements, and asset reliability as examples. The data required covers the spectrum of real-time operational data, environmental monitoring, quality data, and asset information. A connected infrastructure runs from edge connected IoT devices to the cloud, sharing information across individual assets, plants, enterprises, and communities of connected services.

However, the final transformation does not have to be the first implementation. Like any solution, digital tools evolve as our needs and understanding of the problem change. Our use of the data infrastructure will necessarily evolve by:

- Collecting data into the data historian
- Contextualizing an operational "chart of accounts"
- Integrating with external data sources such as LIMS, ERP, CMMS, etc.
- Formulating streaming real-time analytics, first law calculations, and KPI's
- Detecting and notifying on errors and events
- Leveraging complex analytic engines to develop predictive and prescriptive models

Data should be collected once, as close to the source as possible. The data infrastructure will interface with these source systems. As part of the transformation effort, organizations evaluate missing data for the potential value of the information against the expense of collection. Where possible, organizations can leverage IoT technologies or collect manually.

Demonstrated use cases include monitoring site heat index to prevent heat stress, extrapolating corrosion curves to facilitate mechanical integrity programs, performing both golden batch and multivariate analyses, and collecting asset utilization data for reporting among others. Successful digital transformations provide competitive advantages for these enterprises. We believe people with data can transform our world.

# 590961 How to Realize Your Digital Transformation

**Abstract:**

Today, few companies doubt the benefits of Industry 4.0. It is relatively easy to talk about the need for a digital transformation from a high level, but when it comes to planning and executing a specific digital transformation roadmap, it tends to be more challenging. But it doesn't have to be that way. Industry 4.0 has now been around for a few years, which has allowed tried and tested approaches and methodologies to emerge that can be applied to pretty much any size company in any industry.

We will look at the value chain of chemical processing plants identifying application areas for Industry 4.0 technologies such as digital twins. The focus of this industry specific approach is on structured and open data models that break down stand-alone and isolated knowledge silos - in effect integrating the industry relevant value chains. Actionable items include, but is not limited to: the use of multiple digital twins as controlled copies handling business critical information for data-based business management; the continuous assessment of the correct technologies and technology levels to be implemented for each business relevant area; business and site level consulting on industry 4.0 until maturity is reached and digitalization is business as usual in the chemical processing industries

In addition, we will gain insights into how to shape your own future by combining novel and mature technologies in new and exciting ways, which includes project prioritization. It is a critical exercise for companies in formulating a digital transformation because it enables them to identify areas that will generate the greatest value and evaluate each of them based on specific criteria. These can then be prioritized, which will help decision-making and effective resource allocation. This may sound like a simple step, but it requires dialog, involvement from several stakeholders and a framework to put it all together. Industry 4.0 is *not* a one-size-fits-all approach.

By attending this presentation, you will gain the following insights:

- Gain high-level insight into the concept of 4.0 technologies applied in chemical processing industries.
- Gain site-level insight into digitalization use cases transforming our industry now!
- Basic understanding of how to identify Industry 4.0 projects.

View of how to best prioritize them in terms of resources, ROI, digital maturity level and more.

# 589365 Piloting Digitalization at Dupont

**Abstract:**

Okay, you've decided that it's critically important to begin going down the digitalization path. Now what? This presentation will cover DuPont's efforts to date to implement pilot digitalization projects making use of data from existing assets and provide some learnings that you may benefit from in your own piloting efforts.

Specific questions that will be answered and topics that will be covered include...

- Implementation in a production facility making use of real world data
- Selection of digitalization options to be piloted
- Who was involved in the implementation process?
- How long did it take to complete the piloting?
- What support did you receive during the piloting?
- What specific digitalization tools/apps were piloted?
- What was the outcome of the pilot testing?
- What would DuPont have done differently?
- What are the next steps?

# 591511 Electrification, Process Intensification, and Artificial Intelligence for Modular and Improved Chemical Manufacturing

**Abstract:**

There is renewed interest in the chemistry and processing of small molecules, such as carbon dioxide, $CO_2$, ammonia, $NH_3$, and small hydrocarbons (methane, ethane and propane). The large energy demand and associated large production of $CO_2$ in ammonia production, the conversion of $CO_2$ to chemicals and fuels, and the revolution of shale gas in the U.S., and thus of small hydrocarbons, are changing the global energy and chemicals terrain and are creating new opportunities for research and manufacturing. Common to the small molecules are the challenges of activating chemical bonds, the large energy demand, the equilibrium limitations, the loss of selectivity, and catalyst deactivation. In addition, processing some of these resources may require distributed manufacturing, whose economic viability is uncertain. These challenges require understanding of the fundamental chemistry and development of new catalysts and processes. This talk will overview these emerging trends and present examples from chemical activation using electrification, such as microwaves, coupled with other process intensification concepts to allow scalability. The role of data science in process design will also be discussed.

# 590197 Advanced Integrated Techniques for Quantifying Kinetics and Transport Parameters in Multiphase Reactors

**Abstract:**

Multiphase reactors and multiphase flow systems have found extensive related to energy, environmental, chemicals, petroleum and variety of products processes. Multiphase flow reactors are complex due to the intricate interactions and reactions among difference phases. Therefore, sophisticated techniques that integrate measurements in a novel way of various hydrodynamics and transport parameters and their integration with kinetics are required to quantify and visualize these systems to provide benchmarking data for validating CFD and models. These advances should provide the mechanistic frame work to properly scale up the results from lab to commercial scales.

In our laboratory, various types of multiphase systems such as slurry/basket reactors and packed bed reactors under high temperature and high pressure for various industrial and environmental processes have been studied by developing advanced measurement techniques and validating CFD simulations which enable process intensification and facilitating the implementation of new mechanistic scale-up methodologies. In general, these techniques can be classified into two categories: radioisotope techniques and non-radioisotopes techniques. The radioisotope techniques include radioactive particle tracking (RPT), dual source gamma ray tomography (DSCT), gamma ray densitometry (GRD) for 3D flow field, velocity and turbulent parameters, phases distribution and flow pattern identification measurements. The non-radioisotope techniques include 4-point optical probe for bubble dynamics, optical probes for gas-solid dynamics, optical probe for local mass transfer, optical probe for local liquid and gas velocities, holdup and their time series distribution in packed beds, gas tracer technique for global mass transfer, heat transfer probe, pressure transducers, and others.

# 591298 Process Intensified Conversion of Stranded Energy Resources to Value-Added Liquid Products for Distributed Production

**Abstract:**

Over 200 billion $ft^3$/year of natural gas at remote production sites is flared in the U.S. due to the limitations in pipeline transportation capacity. This wastes valuable resources, increases air emissions, and burns profits. Similarly, renewable electricity (wind and solar) in some disadvantaged locations becomes stranded due to the limitations in power transmission capacity and intermittent nature of the renewable electivity. This presentation introduces the applications of microwave catalysis for the conversion of these stranded energy resources into liquid transportable chemicals such as aromatics and ammonia. Microwave irradiation has shown a profound impact on catalyzed and uncatalyzed gas-solid reactions. In particular, it has been demonstrated that microwave-specific effects can manifest themselves through the enhancement of reaction rates, changes in the position of equilibria and the distribution of products.

Methane must be activated to a high-energy transition state before it can further react to produce stable products. The high reaction temperatures (>700°C) required to activate methane's strong C-H bonds usually results in rapid and extensive coke formation that blocks catalyst micropores, deactivating the catalyst. The gaps in scientific understanding of the direct, non-oxidative methane conversion process preclude the design of highly selective catalysts and reactors to make desired higher value products. Technical hurdles include catalyst stability and thermodynamic limitations that constrain the desired product yields. In this study, methane conversion reaction is enhanced with microwave irradiation to demonstrate microwave reaction chemistry at catalyst-reactant interface for selective C-H bond activation

Stranded electricity can be converted to liquid ammonia under ambient pressure and 320°C using water and air as feedstock. Conventional ammonia synthesis process, i.e. Haber-Bosch process, cannot be scaled down economically due to the high temperature and high pressure operation and the intermittent nature of renewable electricity that requires frequent startup and shutdown. Catalytic ammonia synthesis was conducted under two microwave irradiation scenarios: microwave and microwave plasma. Other than selective activation of dinitrogen to metastable radicals, the most obvious advantage that microwave irradiation affords in driving a heterogeneously catalyzed reaction is the ability to locally heat the catalytic sites. Many industrial processes utilizing heterogeneous catalysts are high-temperature processes wherein both components of the reaction (i.e., catalyst and medium) are heated to the temperature required for the reaction to occur. Specifically discussed in this presentation are the effects of electromagnetic properties of catalysts, microwave frequency and microwave energy absorbed on the conversion of nitrogen and ammonia yield.

Technoeconomic analysis was carried out to compare microwave catalytic processes with the state-of-the art. The comparisons are made to focus on process intensification elements such as energy efficiency, energy productivity and capital cost reduction. Sensitivity analysis was made to evaluate various scenario and different plant sizes. The resulting engineering innovation includes plant siting flexibility due to smaller scale, lower capital and operating costs, higher energy efficiency, lower carbon footprint, no or low environmental emissions.

The potential for a dramatic transformation in the chemical process industries is emerging from the rapid development of process intensification. This can potentially yield substantial economic and environmental benefits in combination with advanced manufacturing by significantly reducing the size of chemical process plants. This presentation aims to revolutionize the chemical process industries by enabling the development of intensified and, where appropriate, modular process plants.

# 583338 Keynote: Industrial Cleaning: towards condition-based cleaning

**Abstract:**

Cleaning of industrial equipment creates risks, both for the persons who perform the cleaning activity and for the equipment that is being cleaned. Often specific knowledge to failure modus are unknown by the persons performing the cleaning and too often heat exchangers are being damaged due to the lack of knowledge and experience of the people making the cleaning decisions. This keynote addresses some basic guidelines towards the following encountered failure modes: e.g. corrosion, erosion, overheating, incompatible cleaning chemicals. Furthermore personal safety items will be discussed.

# 589781 Condition-Based Cleaning Models for Industrial Heat Exchangers

**Abstract:**

Periodic cleaning of industrial equipment is often necessary to maintain plant performance over long operating periods. Although cleaning decisions (*i.e.* which heat exchanger to clean, when to clean it and – crucially – how to clean it) should be based on economic drivers (including trade-offs between operating costs, throughput of the installation and quality of the cleaning preformed), they are very often made based on operators' experience and simplistic – often misleading – performance indicators (e.g. fouling resistance, pressure drops). Moreover, operators may not be aware of which cleaning technologies are available to them and which is the optimal option to use .

In this paper, a framework integrating condition-based cleaning models that capture kinetics and efficiency of a cleaning action as a function of the characteristics of fouling deposits (e.g. composition, layering, degree of coking) is presented. The approach is used to provide actionable information on the best method to be used, duration of the cleaning and expected economic benefits, depending on actual equipment conditions. Results are presented for individual heat exchangers and entire networks.

# 588894 Monitoring-Based Fouling Mitigation in Crude Unit Preheat Trains Using Lab and Field Fouling Units

**Abstract:**

The physical and chemical mechanisms associated with crude-oil fouling are complex, which hinder effective applications of mitigation methods for the preheat-train (PHT) exchangers. In order to develop an effective mitigation strategy, it is essential to perform a root-cause analysis (RCA) to identify the source(s) of fouling precursors. The next step is to evaluate ways to interrupt or eliminate the source of fouling precursors and conditions under which they are generated.

The multi-year investigation in this presentation focused on monitoring-based mitigation to systematically perform the root-cause analysis using lab and field fouling units. The analysis included the effects of crude-oil blending, physical parameters (such as flow velocity and wall shear stress), wall temperature (heat flux), performance of desalter, iron-sulfide induced fouling, asphaltene conversion and the effects of inorganic salts. The fouling data obtained with lab and field-fouling units were used to develop the threshold fouling model, which was then used to develop maps that help identify PHT exchangers that are in the high-fouling zone. The threshold models were also implemented in Hexxcell Studio™ to assess the economic impact of fouling and predict future performance in the pre-heat train operation.

The combination of the fouling data from the lab and field fouling units and the simulations were used for the root-cause analysis based on which an effective mitigation strategy can be developed and its benefits assessed.

# 586683 Advanced Process Control Optimum Maintenance

**Abstract:**

Several articles report that after an 18-24 month period over half of the Advanced Process Control (APC) installations are performing at either pre-installation levels or have been removed. There are diverse reasons for the performance shortfall; a major one is that the process characteristics drift from those that generated the controller model. The core of APC is the controller model, typically a linear local representation of the process (gains and dynamics), which is obtained by best matching the model to process responses from incremental changes in the manipulated and disturbance variables. But, when throughput rates and other attributes of the process (reactivity, efficiency, fouling, feed components, etc.) change, the model no longer represents the process, control degrades, and the benefit of the APC diminishes. In this case, recalibration of the model (retesting the process response and adjusting model coefficients to best match) can return the APC to full functionality.

The benefits of APC include increased throughput, reduced variation, constraint avoidance, etc. And these can be combined to represent a daily economic-equivalent benefit for the application, the same anticipated benefit that led to justification of the APC installation. As the process characteristics deviate from those representing the model, control functionality degrades, and the equivalent economic benefit of the APC installation declines. Recalibration has a cost, which is desirably avoided. Recalibration of the APC can restore APC benefit. If recalibration is performed when the functional benefit of the APC installation is high, there is little gained from the recalibration cost. However, if recalibration is postponed, the loss of economic benefit of the APC installation can be greater than the cost of recalibration.

A procedure has been developed that employs dimensionless equations to compute the optimum recalibration interval. This optimum recalibration interval is based on a form of decline in performance that is modeled based on the owner-user performance criteria. Further, the dimensionless numbers can be used to compare the performance across units of similar types and across industries. The form of the equation allows for quantitative processes to change the interval for continuous improvement in the recalibration procedure. This continuous improvement process is based on learning curve theory.

# 590698 Benefits of Deploying PACE, a New APC Technology

**Abstract:**

Advanced Process Control, specifically model based multivariable control, has been applied to air separation units (ASU's) since the technology was introduced to the market in the mid 80's. ASU's have presented many challenges to traditional APC. A new generation of control technology was released in 2015 between Shell and Yokogawa called Platform for Advanced Control and Estimation (PACE) that helps effectively deal with many of these challenges. This paper will explore those challenges and describe the features of PACE that are used to overcome them and drive more benefits from each application.

# 590362 Batch APC: Addressing a Long-Standing Problem in the Process Industry

**Abstract:**

*Advanced Process Control has been a reality in the continuous process industry for over 25 years. However, the application of advanced control technologies to the Batch industry has been scattered and mostly concentrated on the Batch monitoring. There are many challenges involved in controlling and optimizing a Batch process such as nonlinear time-varying, quality variables available only after completion of batch and manipulated variables have very different effects at different times during the batch. In this context, there is an opportunity to better control product qualities at specified targets and optimize batch cycle times by implementing an advanced control strategy that can predict final product quality and optimize decision variables. This presentation will show the latest efforts AspenTech has taken to address this long-standing problem in the process industry to tighten control over final product quality, reduce product quality variation, increase productivity per batch and reduce batch duration.*

# 591336 Process Intensification of Ternary Distillation Using Dynamic Optimization Method and Data-Driven Approach

**Abstract:**

Distillation is the dominant separation technology for liquid mixtures in chemical industry. Studies revealed that over 40,000 distillation columns in operation in the US are responsible for more than 90% of purification needs [1]. The thermodynamic efficiency of distillation can be as low as 10% [2], and the operation of distillation can account for half of the energy consumed by chemical plants [3].

Process intensification (PI) is by now an accepted means for improving the energy efficiency of distillation column (sequences). Through the implementation of innovative design changes, process intensification significantly lowers both capital and operating costs. However, tight design integration/integration comes with control challenges: a reduced number of control degrees of freedom and a more constrained operation window compared to conventional columns [4].

In this paper, we rely on a different approach to process intensification, implemented in the temporal domain, which we refer to as "dynamic process intensification (DPI)." DPI aims to improve process efficiency by exploiting and favorably manipulating in the time domain the nonlinear dynamics of existing columns [5-7]. The principle of DPI consists of imposing operational instead of design changes, aiming to meet product quality constraints while minimizing energy per production rate through implementing a transient (periodic) operating pattern. Thus far, such patterns were defined in terms of switching between a finite (generally, two) fixed operating points. In this paper, we introduce a novel approach based on dynamic optimization. We introduce a data-driven approach for learning the DPI optimization-relevant dynamics of a distillation process from its operational history and present a dynamic optimization-based DPI framework. We show that, among others, Hammerstein-Wiener (HW) models that have fewer parameters to regress and can be trained with a considerably smaller amount of data can accurately capture the dynamics of the distillation plant along with its control system [8-9].

We will apply this conceptual framework to an extensive case study separating cyclohexane, toluene and m-xylene mixture. Meeting all quality constraints, DPI can offer considerable energy savings compared to steady-state operation of the same column.

**References**

[1] Humphrey, J. L. Separation processes: playing a critical role. *Chemical Engineering Progress,* **1995**, 91.na

[2] Humphrey, J. L.; Seibert, A.; Koort, R. Separation Technologies: Advances and Priorities,

tech. rep., Office of Industrial Technologies, Washington DC, **1991**.

[3] Isopescu, R.; Woinaroschy, A.; Draghiciu, L. Energy reduction in a divided wall distillation column. *Revista de Chimie* **2008**, *59*, 812-815.

[4] Schembecker, G.; Tlatlik, S. Process Synthesis for Reactive Separations, *Chem. Eng. Process*, **2003**, *42*, 179-189.

[5] Baldea, M.; Edgar, T. F. Dynamic Process Intensification, Dynamic Process Intensification, *Current Opinion in Chemical Engineering*, **2018**, *22*, 48-53.

[6] Yan, L.; Edgar, T. F.; Baldea, M. Dynamic Process Intensification of Binary Distillation Based on Output Multiplicity, *AIChE J*, **2019**, *65(4),* 1162-1172.

[7] Yan, L.; Edgar, T.; Baldea, M. Dynamic Process Intensification of Binary Distillation via Periodic Operation, *Industrial & Engineering Chemistry Research,* **2019**, *58(15)*, 5830-5837.

[8] Richard C Pattison, Cara R Touretzky, Ted Johansson, Iiro Harjunkoski, and Michael Baldea. Optimal process operations in fast-changing electricity markets: framework for scheduling with low-order dynamic models and an air separation application. *Industrial & Engineering Chemistry Research*, 55(16):4562–4584, 2016

[9] Ian Goodfellow, Yoshua Bengio, and Aaron Courville. Deep learning. *MIT Press*, Cambridge, MA, 2016.

# 589766 Process Intensification of $CO_2$ Absorption Using an Additively Manufactured Intensified Packing Device

**Abstract:**

A multifunctional packing device has been developed, characterized, and tested for process intensification of $CO_2$ absorption by aqueous amines in packed columns. The multifunctional device is an additively manufactured packing element that integrates effective contact of phases and heat exchange in multiphase systems, and can be employed to vent excess heat from an exothermic reactive system or supply thermal energy to an endothermic chemical reaction. The impact of excess heat removal by the device on the absorption of $CO_2$ by aqueous solutions of monoethanolamine is investigated. The design process of the device and the validation of core hydrodynamic metrics will be discussed. Heat and mass transfer testing demonstrated that the device was effective at removing excess heat from the reaction and, furthermore, that cooling could successfully enhance capture of $CO_2$ by the amine solvent. The magnitude of the enhancement was found to depend strongly on operating conditions, in particular flow rates of $CO_2$, air, and solvent concentration. This device represents a successful example of the role of additive manufacturing in the intensification of chemical processes.

# 590385 An Efficient and Reliable Numerical Method for Computing Coupled Multicomponent Fluxes for Complex Mixtures across Asymmetric Membranes

**Abstract:**

**An Efficient and Reliable Numerical Method for Computing Coupled Multicomponent Fluxes for Complex Mixtures Across Asymmetric Membranes**

Dylan Weber[1], Ronita Mathias[1], Ryan Lively[1], Chau-Chyun Chen[2], Joseph Scott[1†]

1 Department of Chemical and Biomolecular Engineering, Georgia Institute of Technology, Atlanta, GA 30332, USA

2 Department of Chemical Engineering, Texas Tech University, Lubbock, TX, 79409, USA.

*Corresponding author; Email: joseph.scott@chbe.gatech.edu*

**Abstract**

Membranes offer a promising means to intensify many industrial separation processes that are currently achieved using distillation and other thermal separation technologies. Historically, key limitations in the design and synthesis of high-performance membrane materials has posed the most significant barrier to the adoption of membrane processes in industry. However, with steady improvements in membrane material properties and module construction, the lack of reliable membrane modeling and simulation tools is emerging as a major barrier in its own right. In particular, widespread industrial adoption of membrane processes as an alternative to conventional thermal separations is unlikely to occur in the absence of (i) reliable predictive models for membrane performance when confronted with complex multicomponent streams, (ii) methods and databases for parameterizing such models using pure component data and limited multicomponent measurements, and (iii) efficient and reliable numerical solution methods that are compatible with standard process flowsheeting software and enable technoeconomic comparison of membrane and non-membrane processes on a consistent basis.

In this presentation, we will describe a novel numerical solution procedure for computing transmembrane fluxes for non-ideal multicomponent mixtures in contact with an asymmetric membrane. Numerous relevant applications of membrane separations such as water purification, carbon capture, hydrogen separation, olefin/paraffin separation, and benzene derivative concentration include process streams that are multicomponent nonideal mixtures. Along with that there are many different types of membrane material being used. From highly selective zeolites, carbon nanotubes, ionic polymers, mixed-matrix MOF and polymeric, graphene-oxide, and specialty polymers. Many of the current approaches for industrial practice adopt the permeability and linear driving force model to describe permeation across the selective layer. These models are applicable in many cases where the sorption is ideal and uncoupled transport can be assumed. However, this explanation of the system when trying to model many nonideal components and interacting permeants is not correct. That is why our group is looking to employ a feasible rigorous solution of the Maxwell-Stefan equations to describe the multicomponent mass transport. Past contributions from experts in these equations such as Rajamani Krishna, Benny Freeman, or Pavel Izák employ approximate solutions by evaluating terms at average compositions, assume the permeate composition is zero to allow for an approximate solution, or use finite differences to approximate the differential terms. These studies are also restricted to a ternary system with two permeates and a membrane phase. No work has been done to have an effective and reliable numerical method for a system with any number of components.

As a brief overview, the Maxwell-Stefan (MS) approach towards mass-transport is based on irreversible thermodynamics by balancing the driving forces any component from a mixture experienced in the form of the chemical potential with the collective friction forces experienced by transporting with other components in the mixture. The Maxwell-Stefan diffusivities can be physically interpreted as the inverse friction or drag coefficients of a binary interaction between two permeating molecules or the permeant and the membrane phase. The main advantages of this transport model compared to the Fickian model is the incorporation of driving forces other than a concentration gradient, but namely MS diffusivities are not dependent on the system and are an intrinsic property of the binary component or membrane pair. Note, these diffusivities can still be functions of temperature, pressure, and composition. The difference is the general applicability from single or binary permeation data to the multicomponent system with many nonidealities accounted for. Then with the nature of many nonidealities, there are many problems in solving such systems of equations. The problems associated with the complete solution of the set of MS equations are the numerical methods associated with solving the matrix formulation of the set of coupled ordinary differential equations. Our research goal is to provide a working rigorous solution method that converges to physically meaningful solutions with only feed stream inputs, a thermodynamic activity model with required parameters, and single component membrane diffusivities (in some cases binary component diffusivities). This will provide a working toolset for the practicing chemical engineer to design these systems without having to be an expert in the underlying transport equations and allow for rapid plant design prototyping and screening of various process units. In this work we consider the Maxwell-Stefan as the general transport model, and the multicomponent Flory-Huggins equation to describe the activity of the permeants in the polymer phase. The interesting note is this contribution is applicable to any membrane system that can be described by the Maxwell-Stefan equations.

# 590118 Ethylene Production Using Oxidative Dehydrogenation: Effects of Alternative Separation Technologies on Process Safety & Economics

**Abstract:**

***Summary***

Previous work from this group utilized a HYSYS simulator to predict the performance of an ethylene oxidative dehydrogenation (ODH) production facility using alternative diluents and separation technology not currently used in the standard industrial process [*1-3*]. An ODH process system has been modeled with ethane being fed along with air into a reactor with simultaneous mixing with pure methane to lower the product gas temperature. The methane feed was later separated from the other product gases and eventually fed into a recycle stream. This process was fairly expensive due to the use of a demethanizer within the process recycle stream. Safety and operational improvements for the ODH process was another aspect of this analysis that yielded interesting results. By the inclusion of storage vessels before and after the ODH reactor, the reactor unit can operate even in the event of an upstream or downstream failure, resulting in less downtime for the process.

The current work investigated the development of alternative separation technologies for the production of ethylene for the oxidative dehydrogenation (ODH) reaction of ethane using a HYSYS simulation with economic and safety analyses. The application of polymer-membrane separation and non-cryogenic distillation alternatives, such as bottom-flash, heat-pump assisted and low-temperature distillation, were simulated and evaluated individually and in-tandem with membrane separation. The effects of alternative diluents, process flammability issues, and economic analysis results will also be discussed.

***Literature Cited:***

1.Maffia GJ, Gaffney AM, Mason OM, Techno-economic analysis of oxidative dehydrogenation options, Top. Catal. (2016), http://dx.doi.org/10.1007/ s11244-016-0677-9 (2016, August 16).

2.Baroi C, Gaffney AM, Fushimi R, Process economics and safety considerations for the oxidative dehydrogenation of ethane using M1 catalyst, Catal. Today (2017), http://dx.doi.org/10.1016/j.cattod.2017.05.041 R (2017, May 11).

3.Gao, Y., Neal, L., Ding, D., Wu, W., Baroi, C., Gaffney, A.M., Li, F., Recent Advances in Intensified Ethylene Production: A Review, ACS Catalysis (2019), http://10.1021/acscatal.9b02922 (2019, August 12).

# 614862 Energy Efficient Separations of Olefins and Paraffins through a membrane II


**Abstract:**

### Update on the Separation of Olefin-Paraffin Mixtures with Optiperm$^{TM}$ Customized Amorphous Fluoropolymer Membranes

Abstract: The separation of light olefins from paraffins is one of the most energy intensive in the world. The state of the art separation method is distillation, a technology that has remained unchanged over decades. Membranes hold the promise of a more modular, energy efficient and capital efficient method for separation. However, they have historically been hampered by low performance and stability.

Compact Membrane Systems has developed a novel membrane material using a silver embedded fluoropolymer material to transport olefins across the membrane. This material has demonstrated over 2 years of longevity in laboratory operation and has shown high selectivity and permeance under realistic operating conditions. Additionally Optiperm membranes were trialed at the Delaware City Refining Company in over 300 days of field operations validating the robustness and capability of the technology in an operating plant environment.

During the past year, CMS has developed and installed a laboratory recycle test systems to allow for the continuous testing of spiral wound modules with areas of ~20 square feet. The recycle test unit enables CMS to test larger scale membrane modules under field operating conditions using actual process streams from end users. Furthermore the recycle rig design incorporates our innovative design for hydrating the process feed within the membrane housing, significantly reducing the complexity of the membrane systems operations.

We will share the latest leanings from our recycle rig testing as well as provide updates on the scale up of our flat sheet spiral wound membrane cartridges. We will also discuss the current state of our hollow fiber membrane configuration development and the progress on our joint project with Dow Chemical through RAPID, a manufacturing institute funded by the Department of Energy. The goal of the work with Dow is to produce a chemical grade propylene from a process feed stream with a propylene concentration of greater than 60%.

CMS will also discuss results on laboratory testing of $C_4$ streams. This includes mixed gas separation of n-butane and 1-butene with permeances between 200-500 GPU and selectivity greater than 20. We will share results for a number of operating conditions and will highlight several exciting $C_4$ separation applications including those with isomer streams. In addition, longevity results will be shared demonstrating the stability of the membrane in $C_4$ separations.

# 591159 When Disasters Happen: A Collaborative Approach Towards the Community's Recovery Process

**Abstract:**

The 2020 Virtual AIChE Spring Meeting & 16th GCPS Wednesday keynote "When Disasters Happen: A Collaborative Approach Towards the Community's Recovery Process" will be presented by Thomas Munoz, Emergency Manager and Homeland Security Director with Texas City. As the Emergency Manager and Homeland Security Director with Texas City, the home of 19 chemical plants, he is responsible for the safety of all its citizens along with managing the crisis communication plan with local industry and major stakeholders.

Safety and disaster planning is vital to the recovery process after a major incident. As key critical infrastructure industries, issues may arise that have not been planned for in advance, such as the recovery from an incident. Developing and maintaining partnerships with local partners in your community will support addressing unforeseen challenges. Recovery efforts include planning with key partners and understanding the phases of recovery, which also consider the mental health and well of everyone involved in the response, as well as the community impacted. In this keynote address a discussion of pre-incident planning considerations and strategies as well as lessons learned/after action improvements post-recovery will be discussed. Additionally, collaboration considerations regarding local resources to promote recovery in the event of a workplace incident will be discussed.

C0-presenter Amanda Groller, MS.Ed., LPC will discuss the mental health impacts of a disaster.

# 591582 Plant of the Future

**Abstract:**

For the "Plant of the Future", there are a lot of promising options. Some of the future technology ranges from automated start-up, operations and shut-downs to Artificial Intelligence (AI) predicting process outputs and machine failures to remote locations being fully controlled due to internet connected sites. Some future changes could be process technological changes, such as electrification and closed loop economies. Another key point that will have to be addressed is ensuring site security and safety: both physical and electronic.

In this panel we will talk about opportunities and early successes and pitfalls and challenges. Topics will cover greenfield builds and brownfield expansions and retrofits.

# 588951 Untapped Advanced Control Capabilities in Modern Distributed Control Systems

**Abstract:**

Modern digital Distributed Control Systems (DCS) have become extremely powerful and capable platforms, even to the point of offering Model Predictive Control (MPC) functions within Level 1 process-connected devices. But, while DCS capabilities have increased substantially since their inception in the 1970s, much of their control capabilities remain untapped. This situation is exemplified by the wide-spread fact that many MPC implementations are designed to overcome regulatory control deficiencies. But such practices represent a gross waste of otherwise valuable resources, especially at the DCS, as well as MPC, level.

Observations made across multiple companies and sites indicates that one of the reasons for the poor utilization of DCS functional capabilities is the lack of investment by Industrial Management into staff training, in addition to the inadequate treatment given to process control in many engineering curriculums. With the above in mind, this case study presentation uses a real-world plant problem and illustrates how judicious use of process understanding and regulatory control DCS functions and capabilities provided the means to optimally operate a vital utility system for a large LNG facility.

# 590098 Towards Tailor-Made Automatically Controlled Compositional Gradient Copolymers

**Abstract:**

The technique called Automatic Continuous Online Monitoring of Polymerization Reactions (ACOMP) allows real time monitoring and control of key parameters of a polymer synthesis. ACOMP is the ideal process control system platform to continuously feed data into a closed loop feedback controller to produce polymers of desired specifications, such as fixed molecular weight and copolymer composition. Because the ACOMP system provides continuous, direct, real time measurements of molecular weight, monomer concentrations, and compositions a controller was developed to operate without recourse to a detailed kinetic model or reactivity ratios, making it theoretically suitable to any system. Copolymers of acrylamide (Am) and sodium styrene sulfonate (SS) were obtained following different trajectories for both molecular weight and composition by free radical polymerization in aqueous medium. This talk focuses on previous successful work using this controller and presents the next steps towards tailor-made compositional gradient copolymers. Through coupling with reversible addition-fragmentation chain-transfer polymerization (RAFT), changes in monomer concentration in the reactor led by the controller are directly reflected on each individual macromolecule, allowing for tailoring the gradient composition of the final product. Gradient copolymers show unique behavior in bulk and in solution due to the gradual change in composition within the polymer chains making them interesting for a wide range of applications, such as damping materials, compatibility enhancers in polymer blends, emulsion stabilizers, drug delivery systems and even photovoltaic units. The use of this in-house built controller with RAFT will allow for producing copolymers with whatever desired compositional gradient distribution in an efficient and straightforward manner.

# 588978 Advanced Process Control Capital Investment

**Abstract:**

Advanced Process Control Capital Investment

This presentation describes the overall process for assessing capital expenditures on Advanced Process Control (APC) projects. Specifically, the one-time expenses for implementing the APC are explored as well as the continuing maintenance costs for keeping the APC at optimum functionality and the impact of variability for the different parameters. APC projects in the process industries have very high returns but have seen a short life. Payouts can be 4 months to 2 years; however, greater than 65% reach zero performance after 18-24 months. This short life can be attributed in large part to not completing the required maintenance to keep the programs in good working order.

This presentation gives guidance on how to estimate the optimum maintenance cycles for these projects, and to include that in estimating the capital cost of the project. Both net present value and internal rate of return are presented as ways these projects can be compared with other competing projects for the funding. Expectations that should be taken before a project is undertaken are defined in this article. The focus is ensuring that the owner/operator will have expectations for maintaining the overall APC going into the project. Additionally, the concept of continuous improvement is introduced so that a plan can be established and budgeted.

# 592114 Process and adsorbent for separating ethanol and associated oxygenates from a biofermentation system

**Abstract:**

Abstract for poster presentation or talk:

Abstract Title: Process and adsorbent for separating ethanol and associated oxygenates from a biofermentation system

**Presenter Information: Deepak Sharma; Sr. Process Engineer, Engineering and Technology, Bayer AG**

Patent number: US8658845B2

Patent Authors:

Deepak Sharma - Sr. Process Engineer, Bayer Ag, St. Louis; (Presenting Author)

Anil Oroskar – CTO, Orochem Technologies, Napervielle, IL;

David W. House – Scientist, Orochem Technologies, Napervielle, IL;

Alice M. Havill – Scientist, Lanzatech, New Zealand

The principles of process intensification if used on existing processes can lead to development and redesign of existing processes to become more sustainable. This presentation will discuss a process and an adsorbent for the separation of ethanol associated oxygenates from a dilute mixture of ethanol and associated oxygenates in water in the presence of organic compounds derived from a biofermentation process. After pretreatment, the separation is carried out in a simulated moving bed adsorption system employing a stationary phase adsorbent comprising fluorinated carbon or modified C18 silica gel selective for the adsorption of ethanol and associated oxygenates, such as 2,3-butanediol, with a mobile phase desorbent selected from the group consisting of methanol, ethanol, propanol, and methyl tertiary butyl ether. The process is useful for removing water from dilute aqueous mixtures of organic compounds comprising ethanol in dilute concentration in water and produced by fermentation, biomass extraction, biocatalytic, and enzymatic processes which are not economically recoverable by conventional distillation methods.

Keywords: Separation; Adsorption; SMB; Purification

Patent Assignee: Orochem Technologies LLC

# 592111 Pump safety, the undiscussed topic in chemical industry leading to several fatal incidents


**Abstract:**

Abstract Title: Pump safety, the undiscussed topic in chemical industry leading to several fatal incidents



**Presenter Information: Deepak Sharma; Sr. Process Engineer, Engineering and Technology, Bayer AG**



Pumps are one of the equipment's in the process industry that require a thorough review during HAZOP studies. The constant source of energy given to the pump, if not converted to kinetic energy to flow the fluid, can lead to heating of the pumped fluid. If the fluid keeps accumulating the energy provided by the motor, in a blocked-in scenario, it can lead to very high temperatures eventually leading to loss of containment of the pump and the pump can get damaged. There have been instances in the industry where fragments of damaged pumps have flown several feet away from the point of loss of containment leading to injuries. The poster will outline a scenario where a pump is run blocked in on suction and discharge for water and a chemical which has a low thermal decomposition temperature. Arc data is used to evaluate the onset temperature of thermal decomposition. The poster will outline safeguards that can be employed to avoid catastrophic loss of containment of the pump. Outlined below are safeguards used to avoid loss of containment of a pump in case of a blocked-in scenario:

- Install PSV on suction of pump
- Employ a low amp/power interlock on the pump to turn off the pump in case of blocked in situation
- Have a high-pressure interlock on the discharge line of the pump to direct flow back to the tank in case of pump deadhead
- Install an Interlock to turn off the pump in case the feed forward valves (automatic valves) close inadvertently

Pump deadhead scenarios are not only catastrophic when in flammable service there are several salts that have low thermal decomposition temperatures. Pumps in water service left blocked in could also blow catastrophically leading to reportable incidents. I feel this is an important topic to bring to the attention of students and professionals in the chemical industry. Poster will include pictures of actual installed pumps along with required safeguards in the field along with supporting P&ID depictions.

# 589440 New Kind of Linear Nanocomposite Applied in Improving Rheological Behavior of Crude Oil and Its Mechanism Studied through Molecular Simulation Method

**Abstract:**

In this work, a new kind of linear nanocomposite produced by carbon nanotube(CNT) as pour point depressant (C2-100) was developed by blending with MCNT nanoparticle and copolymer named Ethylene-vinyl acetate (EVA). Its effect of reducing yield stress and modifying the morphology of wax agglomerate were also evaluated. Besides, the synergetic effect of C2-100 nanocomposite with the magnetic field was also conducted. The mechanism of modifying rheological behavior of waxy crude oil was studied through molecular simulation method, which aims at understanding the interaction between this nanocomposite and wax molecule. The results indicated that C2-100 nanocomposite could reduce the yield stress of waxy crude oil more effectively than the copolymer (EVA) even at low concentration, as well as changing the morphology of wax crystals and agglomerate. Moreover, its effect of reducing yield stress becomes more obvious when the alternate magnetic field was applied. In addition, the results of molecular simulation also presented that the effect of improving rheological behavior of C2-100 nanocomposite is related to the adsorption behavior of wax molecules on the surface of MCNT nanoparticle, which is adsorbed with EVA molecules. In addition, the adsorption behavior of different fragments of EVA molecules can also influence the adsorption behavior of wax molecules on the surface of EVA-attached MCNT nanoparticle, which may influence the effect of C2-100 to improve the rheological behavior of waxy crude oil.

# 589683 Solar Energy Conversion and Noise Characterization Sensing in Building Integrated Solar Photovoltaic Devices with Ventilation

**Abstract:**

A study on solar energy conversion and noise characterization sensing in solar photovoltaic devices with ventilation is conducted. A building integrated parallel plate solar photovoltaic (PV) device was built with a pair of PV modules, a ventilated air cavity and an insulating back panel of plywood board filled with polystyrene installed in an outdoor test-room. Photovoltaic amplification was attained with power output from a potentiometer through rotation of its circular knob. A parallel plate PV device was studied for its
electrical parameters as Resistance-Capacitance (RC) electrical analog circuit. The effect of inductive and capacitive heating losses was considered in evaluating electrical characteristics of a PV device exposed to solar radiation. The characterization of noise interference, due to power difference of two intensities in a wave on PV Device is presented. Noise characterization of a wave is obtained depending on type of wave. The difference of two power intensities is due to transmission of light, sound, heat, electricity, fluid, and fire into a parallel plate PV device. Based on analysis of a solar PV device, a definition with measurement expression and unit for an electrical noise were coined. Standard definitions of noise sources, their measurement equations, their units and their origins under limiting reference conditions were also devised. A review for sensing noise of physical agents is conducted for sensing noise of physical agents (sun, light, sound, heat, fluid, electricity and fire). With integrated environmental control, these environmental parameters are to be characterized through various environmental monitoring sensors.

# 589945 Structural Characterization and Model Representation of Vitrinite-Rich India Coal By Using Advanced Analytical Techniques

**Abstract:**

The detailed investigation of the chemical structure of Karharbari coal (vitritine rich subbituminous coal from India) by utilizing advanced instrumental analytical techniques such as solid state $^{13}C$ nuclear magnetic resonance ($^{13}C$-NMR) spectroscopy, Fourier transform infrared (FT-IR) spectroscopy, X-ray diffraction (XRD), Flash-pyrolysis gas chromatography mass spectroscopy (Py-GC/MS), Laser desorption ionization time-of-flight mass spectroscopy (LD-TOF-MS), X-ray photoelectron spectroscopy (XPS), and high-resolution transmission electron microscopy (HRTEM). The Karharbari coal was an amorphous loosely organized lower rank coal with aromaticity 0.79. This was indicated by XRD results and also confirmed with $^{13}C$-NMR results (aromaticity =0.84) and FTIR results (aromaticity = 0.82). The average molecular weight of the aromatic cluster is 507 amu by NMR result which is closer to the result obtained by HRTEM (MW = 530.46 amu). The presence of different heteroatoms like carbon, oxygen, nitrogen and sulphur with their functionalities, were determined by using a XPS technique. Different carbon/oxygen forms present in the Karharbari coal are (C-C) and (C-H) 68.53%, (C-O) 23.46%, ($COO^-$) 1.96%, (C=O) 6.05% and different nitrogen forms such as pyridine, pyrrolic, quaternary, oxidized and their distribution (mol %) are 19.36%, 45.68%, 31.25%, 3.71 respectively, and different sulphur forms were present i.e. corresponding to thiophenes, sulphones, sulphuric acid and sulphates with the molar content of 16.47%, 41.61%, 21.37%, 20.55% respectively. The construction of the individual molecules will be discussed in detail in the rest of this paper. The individual molecules were geometry-optimized and Materials Studio Amorphous Builder module was used to construct a three dimensional associated structure.

# 589479 Life Management of High Energy Piping with Highlights from ASME B31.1

**Abstract:**

High energy piping (HEP) systems – or as defined by the ASME B31.1 Power Piping Code, Covered Piping Systems (CPS) – are critical assets with respect to production facilities. A failure can have catastrophic consequences from a safety perspective and in terms of equipment loss and downtime. In fact, loss of high-pressure steam due to a failure in a main run of steam supply piping could shut an entire production unit down. Because of this and the requirements of the Code, HEP maintenance and inspection programs are established to ensure that the integrity of the systems is maintained in a safe and reliable condition throughout their lifecycle. There is no one-size-fits-all for the life management of these critical assets because there is a great variety in age, design, materials, and owner objectives. Life management of these assets is beneficial in that it provides a structured, systematic approach to creating and implementing a program including evaluating the current program (whether formal or informal), what the objectives of the program are (short and long term), what equipment is in place, and its history.

A HEP program put in place for overall life management is not synonymous with an inspection program, although inspections are an important part of an overall program. Inspections should be used in conjunction with analytical tools, real time monitoring, and laboratory examinations. The overall approach is life management – not inspection management – of the equipment.

This paper describes a HEP life management approach which includes identifying locations of interest, risk-ranking those locations, implementing a plan for determining the health of the highest risk locations, and taking actions based on the system health that has been determined. This paper also presents and discusses some of the requirements of the B31.1 code, which at the highest level states that each operating company shall develop operation and maintenance procedures for piping systems deemed necessary to ensure safe facility operations based on the provisions of the code, relevant industry experience, the operating company's experience (and knowledge of its facility), and the conditions under which the piping systems are operated.

# 590131 Optimization of C3MR LNG Liquefaction Process Using a Parallelized Simulator

**Abstract:**

Controlling the cost of energy-intensive processes such as C3MR (propane/mixed refrigerant) LNG liquefaction is a prime opportunity for optimization. Our work combines different C3MR models to develop an optimization that uses multiple independent variables, relaxed constraints, and flowsheet alterations to achieve an optimal operating cost. This work uses a rigorous model of the process in a process simulator rather than a reduced model of the process response. Use of a first-principles process simulator increases the accuracy of the model and therefore allows a more realistic optimization, at the cost of increased computing time. Parallelization (using multiple CPU cores) in the process simulator greatly increases the speed of the optimization calculations. CHEMCAD's optimization tools and multi-core technology make it possible to easily define the optimizations, run the optimizations efficiently, and generate global optima with a high degree of confidence.

# 590196 Modifying Fischer-Tropsch Reaction Sequence with Addition of Zeolite to Iron-Based Activated-Carbon Supported Catalyst

**Abstract:**

In addition to previously presented scientific work, modified reaction sequence of Fischer-Tropsch (FT) process was studied by addition of Zeolite ZSM-5 to an iron-based activated-carbon-supported catalyst. Main reactions involved due to FT catalyst are Fischer Tropsch Synthesis (FTS) and Water Gas Shift (WGS) reactions. Main reactions due to ZSM-5 are cracking, oligomerization, isomerization, cyclization, aromatization, conversion of alcohols to hydrocarbons, and coke formation. The waxy and linear structure FT products are converted to gasoline range, lighter hydrocarbons, iso-paraffins, and aromatics using ZSM-5. The catalyst ratio of 1:2, FT catalyst: ZSM-5, is used at 300 PSIG and 300°F. Separate-bed configuration, i.e. FT and ZSM-5 are separated in a reactor, arrangement is used for the reaction sequence analysis. Alcohols with carbon-chain from 3-5 produced by FTS are converted into hydrocarbons over ZSM-5. Waxy hydrocarbons with carbon-chain greater than 23 produced by FTS are cracked and isomerized to small n-paraffins and iso-paraffins using ZSM-5. The hydrocarbon with carbon chain from 7-10 produced by ZSM-5 cracking reactions are converted to aromatic products over ZSM-5. Produced gasoline to kerosene range n-paraffins are further cracked into smaller chain (1-5) iso- and n-paraffins. Some of the iso-paraffins are oligomerized and aromatized over ZSM-5 to form more aromatics. Further, multiple aromatic rings combine to undergo cyclization and dehydrogenation reactions over ZSM-5 to form coke. Water produced as a byproduct of FTS is combined with some of the carbon monoxide, which is the WGS, to form carbon dioxide and hydrogen.

# 590427 Lateral Retention Force Required for Sliding Saline Droplets over Silianized Silicon Surface

**Abstract:**

The lateral retention force required to slide a liquid drop on a surface is expected to increase with the time that the drop rests on the surface prior to sliding till it reaches a plateau. The molecular reorientation of the molecules at the periphery of the solid surface subjected to the unresolved vertical component of surface tension force causes this phenomenon. In this system, the surface tension of deionized water is increased by adding NaCl and is expected that the lateral retention force is increased proportionally with concentration. The retention force increased with increase in concentration reaching a maximum only to drop to a minimum with concentrations closer to saturation. That is, the lowest sliding force is required for both lower and higher concentrations of NaCl. This sudden decrease in observed lateral force is not a function of surface tension and the surface on which the drop slides. This phenomenon is associated with adhesion of salt to the surface and is supported by the drop sliding motion and depinning of the liquid from the solid.

# 590433 Study of Droplet Motion across Polymeric Substrates By Centrifugal Adhesion Balance (CAB)

**Abstract:**

On a given surface, the contour of the drop changes accordingly based on factors of the force applied, chemical structure of the fluid, drop size, and surface topography. With increasing force that is exerted on the drop, we determine the comprehensive trend for drop motion with respect to the controlled factors mentioned above. Drop motion is categorized into four phases and the point at where the initial displacement of receding line is observed is defined as the onset of motion. In this study, the onset of motion is affected by the factors mentioned above and aids in the determination for another value correlation known as Gs, known as Interfacial Modulus. The resistance to interaction towards the droplet as observed is quantified with respect to change in surface topography. This study therefore characterized drop motion and specifically onset motion based on controlled variables (factors). Work of adhesion is also quantified across different substrates to characterize the detachment of liquid from its substrate. All these parameters such as Gs, onset of motion, contact angles were analyzed with different polymeric substrates, drop sizes, and various fluids used.

# 590665 Population Balance Modeling and Spatially-Oriented Control for Crystal Size and Shape in Crystallization Processes

**Abstract:**

Crystallization is an important separation and purification process with relevant applications in the food, pharmaceutical, fine chemical and fertilizing industries. In these processes, modeling and process control for crystallization are focused on crystal size distribution (CSD) and shape of the crystals in order to make it possible to achieve the desired quality and purity of the product of interest. The control of crystal size and shape is of considerable importance for industrial crystallization processes, as crystal morphology has great influence on downstream steps and also from the point of view of product specification criteria. The consideration of crystal shape in modeling of crystallization processes is also important due to its impact on product effectiveness, such as the bioavailability and tablet stability in pharmaceuticals. Control of the crystallization process is can be facilitated with online measurements, which provide information to be used in the modeling of the process. However, if online measurements are not available, the strongly nonlinear dynamics of these processes makes difficult to establish relations between some inputs (e.g. temperature) and the final desirable characteristics of the crystal product. In this context, image analysis has been shown to be a suitable technique to monitoring the process [1].

In our present work, a population balance model-based scheme for controlling the size and shape of crystals is presented. The modeling was performed using a two-dimensional population balance (2D-PBM) in terms of the two characteristic dimensions of the crystals, describing the evolution of crystal size and shape distributions along the batch time. To identify the nucleation, growth and dissolution kinetics parameters using the 2D-PBM approach along the length and width directions, QICPIC-LIXELL's high speed dynamic image analysis equipment was used for the online measurement of the CSD, shape parameters and aspect ratio distribution of crystals during the experiments. In this way, the developed model allowed for the prediction of the crystal size and shape distributions during growth-dissolution cycles.

The developed framework was employed to obtain optimal temperature policies for controlling the crystal size and shape distributions that meet target criteria for size, morphology and productivity using the dynamic image analysis along the runs. Based on the concept of using measurements along batch crystallization processes as a trajectory in a phase space [2], a trajectory-endpoint control problem is presented with an oriented control scheme that is applied to the shape of the crystals. The scheme was used to control the batch cooling crystallization of potassium dihydrogen phosphate. The reported experimental data demonstrate the application of the control policies for temperature to produce crystals of the desired average shape and size in different batch times. Thus, using the first principle modeling coupled to the developed trajectory tracking control, the proposed framework has the potential to be applied to industrial crystallization processes for shape control and to be extended to polymorph control investigations.

References:

[1] Eisenschmidt, H.; Bajcinca, N.; Sundmacher, K. (2016) Optimal Control of Crystal Shapes in Batch Crystallization Experiments by Growth-Dissolution Cycles. *Crystal Growth & Design*, v. 16, p. 3297-2803.

[2] Griffin, D. J., Grover, M. A., Kawajiri, Y., Rousseau, R. W. (2015) Mass-count plots for crystal size control. *Chemical Engineering Science*, v. 137, p. 338-351.

# 590450 Auditing the Process Hazard Analysis - Key Characteristics to Scrutinize in a Good-Quality HAZOP Study and Supporting Effective HAZOP Study Updates

**Abstract:**

Although Process Safety Management (PSM), requiring good-quality Process Hazard Analyses, was promulgated in the United States in 1992, and HAZOP Study techniques date back to the mid-1970's, the HAZOP Study technique is still not uniformly applied in a high quality fashion. The objective of this paper is to provide experience-based tips on Process Hazard Analysis auditing, and some key characteristics that the auditor can scrutinize to assure quality implementation. Analogously, a similar set of characteristics that can facilitate future HAZOP Study updates can also contribute to a quality HAZOP Study that supports the spirit of Safety Management System objectives for the identification of hazards potentially impacting plant personnel, the community, and the environment. This paper will also review tips for structuring the HAZOP Study to facilitate later updates and revalidation.

# 590342 Development of High Performance Organic-Based Cathodes

**Abstract:**

For lithium-based batteries, structure/property relations in terms of cell voltage are desired for new organic-based cathode active materials. In this initial pursuit of structure/property relations, ab-initio calculations were performed on a few organic-based molecules to show the effect on the theoretical cell voltage by inclusion of nitrogen and/or oxygen. Preliminary test results are presented on a newly developed molecule and methods to potentially improve on the cell performance are given.

# 590705 Electrochemical Kinetics of Degradation of Graphite Anodes in Lithium Ion Batteries

**Abstract:**

Growth of the solid electrolyte interphase (SEI) is a major driver of capacity fade in LIBs. Despite its importance, the fundamental mechanisms remain unclear, primarily because of the complicated reaction pathways involved[1–3]. SEI growth can be both electrochemical and chemical in nature[4], and thus, it is a strong function of the potential and degree of lithiation of the electrode. We model the early-stage and long-term growth of SEI by accurately capturing the potential dependence of its formation kinetics as well as long term rate limiting steps, and validating it against the world's largest open source battery cycling data, generated in-house[5]. This is done using the Multiphase Porous Electrode Theory (MPET) framework[6] on graphite (phase separating) and carbon black (non phase separating) particles.

Lithium plating is another key degradation phenomenon that has been elusive, and it becomes important while trying to fast-charge batteries, i.e., 0% - 80% state-of-charge in 30 mins. We show that lithium plating is a key function of electrode morphology, phase-separation dynamics and potential. Phase-separation in graphite is modeled in the electrode using the Cahn-Hilliard Reaction framework described by Bazant[7]. We understand the electrochemistry of the onset of lithium plating with in-situ measurements connected to real time cell potential in a phase-separating electrode for the first time[8].

Results indicate that the peak SEI-forming currents are higher for higher driving currents. Also, we find that SEI only grows during electrode lithiation, i.e. the battery only degrades while being charged. We also find that onset of lithium plating is correctly captured only when phase separation in active material is accounted for. Further, the onset of plating is delayed on electrodes with a thick SEI layer – understanding SEI/plating coupling is integral to predicting fast charging manufacturing protocols for LIBs. This work holds promise for the predictive design of procedures[9] for manufacture and formation of LIBs.

[1] Cohen, Y. S.; Cohen, Y.; Aurbach, D. Micromorphological Studies of Lithium Electrodes in Alkyl Carbonate Solutions Using in Situ Atomic Force Microscopy. J. Phys. Chem. B 2000, 104, 12282–12291. https://doi.org/10.1021/jp002526b.

[2] Horstmann, B.; Single, F.; Latz, A. Review on Multi-Scale Models of Solid-Electrolyte Interphase Formation, Current Opinion in Electrochemistry 13, 62-69 2019.

[3] Nie, M.; Abraham, D. P.; Seo, D. M.; Chen, Y.; Bose, A.; Lucht, B. L. Role of Solution Structure in Solid Electrolyte Interphase Formation on Graphite with LiPF6 in Propylene Carbonate. J. Phys. Chem. C 2013, 117 (48), 25381–25389, https://doi.org/10.1021/jp409765w.

[4] Das, S.; Attia, P. M.; Chueh, W. C.; Bazant, M. Z. Electrochemical Kinetics of SEI Growth on Carbon Black: Part II. Modeling. J. Electrochem. Soc. 2019, 166 (4), E107–E118. https://doi.org/10.1149/2.0241904jes.

[5] Attia, P. M.; Das, S.; Harris, S. J.; Bazant, M. Z.; Chueh, W. C. Electrochemical Kinetics of SEI Growth on Carbon Black: Part I. Experiments. J. Electrochem. Soc. 2019, 166 (4), E97–E106. https://doi.org/10.1149/2.0231904jes.

[6] Smith, R. B.; Bazant, M. Z. Multiphase Porous Electrode Theory, J. Electrochem. Soc. 2017, 164 (11). https://doi.org/10.1149/2.0171711jes.

[7] Bazant, M. Z., Theory of chemical kinetics and charge transfer based on nonequilibrium thermodynamics, *Accounts of Chemical Research*, *46*(5), 1144–1160. https://doi.org/10.1021/ar300145c

[8] T. Gao, Y. Han, S. Das, T. Zhou, D. Fraggedakis, N. Nadkarni, C. N. Yeh, W. Chueh, J. Li, M.Z. Bazant, Interplay of lithium intercalation and plating on graphite using in-situ optical measurements, submitted.

[9] Huang, W.; Attia, P. M.; Wang, H.; Renfrew, S. E.; Jin, N.; Das, S.; Zhang, Z.; Boyle, D. T.; Li, Y.; Bazant, M. Z.; McCloskey, B. D.; Chueh, W. C. and Cui, Y.; Nano Letters 2019 19 (8), 5140-5148. DOI: 10.1021/acs.nanolett.9b01515

# 589787 Model Development of Solid Oxide Fuel Cell Thermal Performance Using Artificial Neural Network

**Abstract:**

NASA's Johnson Space Center has recently begun efforts to eventually integrate air-independent Solid Oxide Fuel Cell systems, with landers that can be propelled by LOX-CH4, for long duration missions. Using landers that utilize such propellants, provides the opportunity to use such systems as a power option, especially since they will be able to process methane into a reactant through fuel reformation. To ensure fuel reformation in the systems, Steam Methane Reformation (SMR) will be employed. Various lead-up activities, such as hardware testing and computational modelling, have been initiated to assist with this developmental effort including the development of dynamic empirical models using artificial neural network. Such a modeling approach has shown to be accurate (R^2>0.9) to predict the thermal and electrical performance for SOFC systems. Hence, such models could be used for system optimization and control design.

# Impact of Spacer Nature on Rheological Properties of Cationic Gemini Surfactant

**Abstract ID: 589650**

**Shams Kalam[1], Muhammad Shahzad Kamal[2], Shirish Patil[1] and S. M. Shakil Hussain[2]**

*(1) KFUPM, Dhahran, Saudi Arabia*

*(2) Center for Integrative Petroleum Research, King Fahd University of Petroleum & Minerals, Dhahran, Saudi Arabia*

Surfactants play a vital role in Chemical Enhanced Oil Recovery (cEOR) techniques. Gemini surfactants are a unique style of surfactants having more water-soluble hydrophilic head and oil-soluble hydrophobic tail units. Both groups are connected by a spacer at or near to the head groups. Polymers are added into the surfactants to increase the sweep efficiency of the flooding processes. This paper presents the rheological properties of Surfactant Polymer (SP) solutions in which two recently synthesized cationic gemini surfactants with different spacer nature are used.

The rheological properties of novel SP solutions were determined using Discovery hybrid rheometer (DHR-3) from TA Instrument. The geometric configuration set for this work was concentric cylinder. All experiments were carried out at 80 °C, while the shear rate for steady shear viscosity measurements ranges from 0.001 to 1000 $s^{-1}$. 0.25 weight % of a polymer FO 4290 SSH was used with different concentrations of newly synthesized gemini surfactants for Rheological measurements. Two hours magnetic stirring and twenty hours retention time were provided to make uniform surfactant polymer solutions.

Both surfactants exhibited good thermal stability in seawater with no precipitation. Shear viscosity and storage modulus were measured as a function of shear rate and angular frequency respectively at different surfactant concentrations ranging from 0 to 0.5 mM, where 0 mM shows pure polymer. We discovered that counter ions also influence the rheological properties significantly. Research outcomes revealed that the novel SP solution with a benzene ring and chlorine counterion produces better performance in comparison with benzene ring and bromine counterion. Moreover, the effect enhances if we increase one more benzene ring in the spacer.

This study reports two recently synthesized SP solutions for cEOR processes in carbonate reservoirs. It presents the effect of counterion on rheological properties and improves the understanding of tuning the surfactant structure in making SP solutions to increase the flooding sweep efficiency.

# 590257 Real Time Energy Management Systems Implementation for Continually Optimized Performance

**Abstract:**

It is well known that energy management is key for the optimal operation of refining and petrochemical sites. Industrial energy systems are constituted by complex utilities networks that are interconnected and have complex interactions that in order to achieve it best performance need to comply with increasing environmental rules, push against operating constraints and adapt to altering power prices and rapidly changing equipment availability.

After the initial installation, management and continuous improvement of these systems present challenges not only from a technological but also from an organizational point of view.

The implementation of an advanced digital Energy Management System (EMS) which includes capabilities for multi-period optimized scheduling, real time optimization and monitoring, deployed either as cloud-based or on-premise solutions that could be acting on open or closed loop is known to provides large benefits.

Critical is the sustainability of the EMS in the ever-evolving energy supply and demand of refining and petrochemical sites. Commercial, asset and organizational factors all change and the EMS needs to adapt to the developing environment of each enterprise meaning the type of implementation, including appropriate management structure, culture, best practices and Key Performance Indicators (KPIs) needs to be continually improved.

The EMS implementation at the large and complex Eastman site in Longview, TX, will be presented as an example. It will provide insight on the initial successes achieved in real time energy management for both, cost optimization and energy related KPIs performance tracking. The case study evolves to discuss how technical asset improvements, once replicated in the EMS, provided alternative operating modes, large additional benefits and sustained optimal

**Andrew Lammers, NN2**

Eastman Longview

**Pablo Montagna*, Cristian Rojas**

KBC Advanced Technologies, Inc

# 589818 Analysis of Fuel Cell System through Project Based Learning for Application of Classical Thermal Fluid Concepts in Mechanical Engineering Laboratory Course

**Abstract:**

A hands-on project involving analysis of air noncircular channels in Fuel Cell to measure the pressure was introduced in a thermal fluid laboratory course. Although the students studied fluid mechanics as a theoretical course, this project gave them the opportunity to build an experimental setup from scratch to help showcase their hands-on skills and to apply their understanding of relevant concepts of fluid sciences. The results of the project included a successful test of a conceptual procedure, designed to measure and calculate pressure drop and flow rates across the anode and cathode sides of a fuel cell. The students enjoyed completing and learning from the project. They also provided very useful feedback on the project by incorporating a functional experimental system for the future laboratory project. This project based learning opportunity allowed students to build their confidence in what they are learning as well as provide a glimpse of what the real world may hold.

# 589438 Measured Versus Calculated Sour Gas Properties

**Abstract:**


**Measured versus calculated sour gas properties**
**Abstract:**
Recent Environmental agencies have restricted standards for sulfur emission from gas chemical processing. Sour gas can extremely impact the delivery and processing by corroding the equipment. Gas properties such as compressibility, density and viscosity are a major factor controlling prediction of future production pipelines, and production tubing and pipelines in gas industry.
Sour gas effect on all gas industry aspects usually gas properties are measured in laboratories but occasionally experimental values are not available due to complexity and time consuming hence a reliable method would be proposed.
We are assessing the accuracy of the industry standard methods in calculating sour gas properties, compressibility factor using the following correlations:
Drunchuck - Abukkasem, Hall – Yarbarough, Drunchuck- Purvis- Robinson, Londono et-al, Alanazi et-al and Azizi et-al.
Viscosity using empirical correlations such as Carr-Kobayashi Burrows, Lea-Gonzalez –Eakin and Dean- Stiel.
Density using equations of state Redlich Kwang and Soave -Redlick - Kwang.
This study shows that the best correlation for calculating compressibility factor and density of high concentration sour gas is Londono-et al, all other considered methods are not accurate. All other methods investigated include high error especially for high $H_2S$ concentration.
However the error in estimation of Z factor for most of the considered methods has great impact on estimated sour gas density. Therefore we recommend that sour gas properties be measured experimentally at high pressure and high temperature conditions.

# 587967 The Study of a S-CO2 Oxy-Fuel Combustion Power Plant for Low Grade Coals

**Abstract:**

Recently, it is very important to have reliable solutions that can maximize power generation efficiency of current power generation processes and decrease emissions of CO2 and other pollutants because there are increasing concerns about the global warming and the depletion of fossil fuels. Among the available carbon capture technologies and power generation technologies, a S-CO2 (supercritical carbon dioxide) cycle based oxy-fuel combustion process appears to be a promising mid-term solution for the production of electricity from fossil fuels. Since supercritical carbon dioxide (S-CO2) involves a rapid density change at small temperature and pressure ranges, the S-CO2 cycle can extract large amounts of energy from relatively small equipment and it is possible to minimize or simplify a power generation block and to obtain higher thermal efficiency than existing steam cycles used in power generation plants. Therefore, an indirect S-CO2 oxy-fuel combustion power plants for low grade coals was designed and analyzed by using ASPEN-PLUS. Also, the operating characteristics of the indirect S-CO2 Oxy-fuel combustion power plant were compared with those of conventional air combustion power plants.

# 590099 Characterization and Performance Assessment of Oil Entrapment By Commercially Available Surfactants

**Abstract:**

Oil spills occur frequently and are a threat to the health and safety of human society and the well-being of the environment and ecosystem. Remediating oil spills is an ongoing challenge. Surfactant based dispersing agents, such as the Corexit® product line, are widely used for remediation of oil spills mainly through direct application on top of the oil slick. This work encompasses the investigation of variables that influence the emulsification process and therefore the encapsulation of oil within the surfactant micelles. Of central importance are the conditions under which oil entrapment in emulsions even occurs. A central finding is that there appears to be a threshold shear rate for emulsifying oil with surfactants, below which no emulsification occurs. Different schemes for applying shear stress in this work included simply stirring, using a high energy colloid mill, and flowing the mixture through a high shear capillary. Also, the sequence in which the oil/surfactant/water (O/S/W) components are mixed impinges strongly on whether and how much oil entrapment occurs. To try to correlate shear stress and velocity profiles throughout the sample volume with entrapment efficiency and kinetic stability of these O/S/W systems, Computational fluid dynamics (CFD) simulations were conducted. Another important factor is the ionic strength of the aqueous medium, which greatly impacts kinetics of these emulsions, but with a smaller impact on the threshold at which any emulsion is obtained at all. Surfactants tested include Corexit® EC9500A, sodium dodecyl sulfate (SDS), Triton X-100, and cetyltrimethyl ammonium bromide (CTAB). The quantification of oil entrapment was performed through either thermal gravimetric analysis (TGA) or differential scanning calorimetry (DSC).

# Designing a Native Model for Assessment of the Effectiveness of Advertising

Sorush Niknamian
PhD, military Medicine Department, Liberty University, USA
Email: so.niknamian@gmail.com

## Abstract

Reviewing and evaluating the effectiveness of advertising can be used as a criterion to convert weaknesses into strengths and budgeting. Also, Identification of the factors influencing the effectiveness of advertising can lead to more professional design and implementation. In the banking industry, with the increasing number of banks and increased right of customers to choose, the role of advertising has become increasingly important to succeed in this field of competition; so it should be said that advertising can make the bank close to its goals when it is sufficiently effective. The current research aimed to identify and rank the effective factors on the design of a native model for evaluation of the effectiveness of advertising of the Eghtesad Novin bank in Iran. The statistical population of this research is all managers and experts of the Eghtesad Novin bank. According to the research literature, the criteria and sub-criteria affecting the effectiveness of advertising have been extracted. By using the fuzzy Delphi technique and based on the experts' opinion, the final indices were extracted. Then, using the fuzzy AHP method, the rank of criteria has been addressed. The derived criteria are demographic characteristics, individual emotions, and attitudes of the customer, apparent features of the message, message content, market characteristics, cultural, economic features, design and nature of the service, advertising methods. The results of the ranking showed that the methods of advertisement, the design, and nature of the service, the apparent characteristics of the message, the emotions and attitudes of the customer, the demographic characteristics, the content of the message, economic, market characteristics, were placed in the first to ninth ranks, respectively.

***Keywords***: banking services, banking services marketing, advertising, advertising effectiveness,

**Introduction**:

Advertising is one of the promotional mix elements of marketing mixes, which play an important role in attracting customers. Advertising is a bridge between the organization and customers and can introduce the organization and its products to customers, inform consumers about the goods and services of the organization, convince them about the benefits of the product, and ensure them to have the best choice even after the purchase when the customer is uncertain (Buck, 2016: 172).

Given the increasing competition among the brands in the marketing sector, advertising is a necessary tool to introduce the characteristics of goods and services to customers. Leong et al. (1999) pointed out the fact that consumers overcome advertising information, and many of them do not remember the advertisements that they are exposed to, and it is necessary to consider the advertising and its attractive methods to attract customers' attention. The ability to attract customers' attention is one of the main indicators of advertising effectiveness that should be taken into consideration. The attention of customers is a key component of the overall effectiveness of advertising (Pileliene et al., 2017: 2). Any professional marketing without advertising will not be effective. Despite relatively high advertising costs, advertising is mentioned as an investment rather than the cost. Since many advertisements cause that the advertising companies do not benefit their advertising because of the wrong way of advertising, the competitive companies benefit from the advertising of that company (Hartnet, 2016: 23)

Advertising should be in line with the goals and plans, as well as the strategies of the organization. Therefore, we should see to what extent the advertising has met such goals. Allocating funds without evaluating the effectiveness of past advertising is the waste of resources of the organization. By evaluating the effectiveness of advertising, changes can be made to the form, the content of messages, type of media, budget, communication channels, and even the time and conditions for conducting advertising to create more useful and effective advertising than the past (Yazgirudo-Yasteh, 2015: 359). In the current competitive world, the importance of advertising is increasing. As a result, businesses have devoted a great deal of money to customers' attraction, but it is important to see whether the company advertising has been designed and implemented in such a way to meet the goals of the company. This issue has been often neglected in our country. (Kim, 2014: 2107).

The amazing effect of advertising on the lives of individuals requires that this scientific, social and communicative branch be carefully studied and advertising is made with a scientific, artistic, specialized and expert view. Professional advertisement can be a sharp sword for sales promotion, both inside and outside the country, and otherwise, it is a waste of time, money and energy. A professional advertising lever can, with the help of new technology, penetrate humans, and cause more desire for consumption (Arribeh, 2016: 463). Today, banks have turned their attention to the widespread use of mass media advertising, given the highly competitive structure governing on them. Also, given the relationship among a wide range of businesses of organizations and households and banks through various deposit accounts, loans, and other financial services, banks can benefit from advertising in a proper way to meet their goals, namely the introduction of their services, the communication with customers, their effectiveness on them in choosing a bank to do their banking business and ultimately attracting them. Unfortunately, in most banks, despite the massive advertising and the allocation of large amounts of energy and budget, the effectiveness of such advertisements is less evaluated. Since advertising is one-way, it is necessary to receive feedback from the customers (Rabiei, 2011: 19).

Regarding marketing, the effectiveness of advertising depends on the wide range of coverage of various consumers and the cost of advertising campaigns. Economists have defined the effectiveness of advertising as a change in the profit derived from promotional activities. Therefore, the effectiveness of bank advertising is the potential quality of information reported to consumers, and in promotional campaigns, it aims at creating a bank image and informing banking products and services, as well as obtaining a specific economic result before and after the transmission of the message (Denis et al., 2012: 85).

Eghtesad Novin Bank (EN bank), which is one of the oldest banks in Iran, due to the addition of new banks, should use new tools to protect and even increase its market share for the assessment of the effectiveness of its advertising. For this purpose and given that the bank uses various advertisements, it is necessary to design a model to evaluate the effectiveness of the bank advertising; therefore, the present study seeks to answer the following questions:

- What indicators are used to assess the bank native advertising model?

- How are the indicators for bank advertising evaluation weighed and ranked?

In this research, we have tried to identify the criteria using the Fuzzy Delphi process; then we will use the Fuzzy AHP method to rank the identified criteria. In the next section of the research, the literature has been studied. Then the research design has been expressed, and in the last section the conclusion has been presented.

**1-Research literature:**

**1-1-Marketing of banking services**

The growing trend in service activities has led to the emergence of new markets and large service companies. The complexity of service industries, such as information systems, management consultancy, and logistics, has increased, as service providers have entered the market with a variety of business backgrounds and capabilities to provide a broader range of services. Facing this complexity requires knowledge and skills to understand the services. The definition of the service includes all activities that have been done (for example, writing document specifications) (Selviaridis et al., 2013: 1398).

In the industrial literature, there are various ways to define the service and product requirements. Van Weele (2010) has distinguished technical and applied definitions from each other. The first requirement expresses the exact technical specifications of the goods with the required services and inputs while it focuses on performance demands and demand-oriented performance orientations. In this regard, Axelsson and Wynstra (2002) have proposed four methods for definition of services: input definitions are based on the sources and capabilities of the service provider while the process-oriented approach focuses on how services are presented regarding activities and processes. Output-oriented definitions focus on service capabilities and performance levels, as well as the interaction between the suppliers and recipients. Ultimately, they focus on stress-oriented results in the economic value of services (Selviaridis et al., 2013: 1399).

The financial sector forms an essential part of the service sector. In this regard, banks have used marketing to improve the strategic performance of banks. In particular, since the 1980s, this trend has been confirmed and criticized. Marketing researchers and experts are interested in researching the various aspects of marketing in banking. The quality of services is essential for customers who use banking services, and the provision of accurate and high-quality services leads to the survival of banks in the competitive market. Improving the quality of services will increase the access and response to the demands of the bank's customers, which will make the bank more distinct than other competitors. Therefore, measuring the quality of services in the banking industry is essential. Providing quality services is a hard and difficult task, in this regard, the features of services in general and banking services, in particular, should be explained. Banking services require a high degree of appropriate communication between service providers and customers. If services are not provided properly, they will be at high risk.

About the five features that distinguish them from material services (intangibility, non-allocation of materials for production, diversification, non-corruption, and lack of ownership), financial services have two other specific features, that have been firstly developed by McKechnie (1997). These characteristics include trusteeship and two-way information. These special features mean that the evaluation of services is more difficult than material products, and they not only create a challenge in the development of marketing activities, but also they are the source of the problem in providing an optimal level of quality services. Banking services require a high level of communication between service providers and customers, as well as a high degree of customer participation in the process of providing services. Angurr et al. (1999) emphasized that due to the high standardization and the development of technologies in the banking sector, human factor has been considered as the most important one in making distinctions among competitors. The expertise and motivation of the employees, as well as the level of customer involvement in the provision of services, significantly affect the quality of services. Due to the wide range of banking activities, attention should be paid to cultural and social changes that affect the attitudes of customers (Durdana et al., 2015: 100).

Marketing of banking services has three components as follows: if any of them were properly executed, banking services would change the country.

1- Market study: Identification of the banking services market system, including the status of current banking services and their weaknesses, strengths, expectations, demands, ideas and opinions of customers, the conditions of environmental instruments affecting the market system, in particular government policies and governing laws, and even global changes in the banking system and successful models in this system. Indeed, market study is the foundation of the evolution of the banking system, because market study (which is the result of constant communication with the market and conducting marketing studies and research in various forms) leads to comprehensive, up-to-date, relevant and necessary information for decision-makers and bank managers (Munsung Rheea & Satish Mehrab, 2006: 508).

2. Market creation: It involves the use of a variety of simple and complex marketing techniques and competitive strategies, as well as the growth, development and market creation in the banking system; it needs to understand the internal customers (employees) and external

customers, to communicate with them and meet their needs and demands. Market creation in the competitive environment requires appropriate strategies (Moch, 2013: 2908).

3. Market protection: One of the most important tasks of marketing in the banking system of the country is to maintain market share, a stable relationship with customers, and to convert ordinary customers to loyal and supportive ones. The more intense the competition, the more valuable the loyal customers will be, as marketing research in the world has shown that the cost of finding a new customer in firms is 3 to 33 times higher than the cost of maintaining current customers regarding the type of products and services. Therefore, market protection prevents the costly advertising, and attraction of new customers and provides a kind of confidence and serenity for market survival, especially in competitive conditions (Fredriksson & Moro, 2014: 69).

### 1-2-Effectiveness of bank advertising

One of the indicators of success in any organization is the extent to which the organization achieves its goals. Here, the success of any organization is determined by its own goals. This approach has been called effectiveness in the management literature. The effectiveness is the degree to which the organization achieves its goals. Goals must first be defined to measure effectiveness in any organization. Then, plans are set up to fulfill these goals, and thus the plans will be implemented. In the end, the results are compared with predetermined goals (Manjegani et al., 2015: 124).

About 100 years ago, William Heketh statement, "Half of the advertisements were wasted, we just do not know which half it is" attracted everyone to the accountability of advertising. What matters to advertisers is that how much the amount of money spent on advertising has returned? Has the cost spent on making and broadcasting the advertising been effective? Has the effect of advertising been strong or weak? Knowing the answer to these questions is important to the advertiser because today companies pay much money to make and broadcast advertising. According to Jack Trout, the money we use to advertise is very much. Every second of the World War II was $ 9,000, and Vietnam cost $ 22,000 per minute. But in the American Football Final, every minute of advertising cost $ 2 million, so the advertiser has the right to know about the effectiveness of his advertising (Aksoy, Atilla, 2005, Yeni Reklamcilik, Istanbul Bilgi Universitesi, 95). It is very difficult to measure the impact of long-term advertising on many occasions. Also, academically in the world of advertising, it is impossible to create a comprehensive theory that can be used to determine the effectiveness of advertisements on people at different levels under different circumstances. The effectiveness of media is its capacity to create desired results, or special conditions in a particular position and content, that is, attaining the desired communication goals (Wery, 2012)

Evaluating the effectiveness of advertising indicates how much advertising has succeeded in achieving the goals. The evaluation of the effectiveness of advertising acts as a feedback, informs us about the weaknesses and strengths of our advertising by which we find our shortcomings in the advertising. The marketer may choose an appropriate stage regarding product life or brand position, and use it to define his advertising goals and ultimately evaluate its effectiveness (Eckler & Bolis, 2011: 2).

The effectiveness of advertising is a function of its content (message), the execution (how it is transmitted), and repetition (the number of times each consumer sees the message) (Cutler, 1997, quoted by Moorthy & Hawkins, 2005). Some of the effectiveness criteria are the call and recognition of the brand name (Park, 2006), awareness, willingness and intention to purchase (Decrop, 2007), attitude towards the ad, attitude toward the brand name and intent to purchase (Muda et al., 2014)

The control of the communication element has a direct effect on customer perception of different dimensions of the brand name, such as service quality and loyalty. It means that the elements of communication, advertising and promotion must be carefully implemented and controlled (Suvittawat, 2015). Informative advertising is one of the factors that shape the behavior of the bank customers. Studies conducted both on attracting, retaining and loyalty of customers, show the need to examine the impact of advertising on attracting and loyalty of customers (Ioannou, Boukas and Skoufari, 2014: 25).

An ideal model of world-class advertising can: 1. transfer a consistent meaning to the consumer all over the world. 2. Transfer positive global capacity for the brand. That is, the results of advertising among different peoples should produce a positive effect and transfer high-quality signals (Strebinger et al., 2018: 410).

Advertisers want to prepare consumers to like goods or services, but sometimes it harms them through unwanted and intrusive advertising. It leads to the creation of a series of negative emotions in advertising. The purpose of ads is to create positive feelings in consumers about advertised goods and services (Bell & Buchner, 2018: 1).

On the other hand, research shows that commercial advertising can be positively evaluated and considered, and this positive evaluation leads to a positive evaluation of the brand with the product. However, advertising can have negative effects that lead to the formation of negative attitudes towards that product and service. Therefore, by increasing the volume of messages, it can have positive, negative or neutral effects. Therefore, marketers should evaluate advertising messages and their behavioral outcomes. With this regard, the effectiveness of advertising, based on a hierarchical approach, should be fully investigated (Tafani et al., 2018: 125).

There are various factors which influence effectiveness of advertisement like media, time, market life stage, advertisement life stage (new or old advertisement) and also advertisement creative signs (like content and administrative elements) ( Talis et.al, 2005). Also using popular cartoon personages (Ranjbarian, 2009), size (Rousat, Venous & Ebrahimi, 2004, Baltas, 2003), color (Park, 2008, Moriarty, Mitchell, Wells, 2009, Cheng & Kao, 2011), complexity or simplicity of the advertisement (Morrison & Dainoff, 1972, Di Res, 2008), using irony factor (Mukherjee, Hober, 2004), using famous artists (Rousta, Venous, Ebrahimi , 2004) and consumer features (Matrin, Phaymi, Agee, 2002, Park, 2008) are among these variables. Advertisement repetition is also a factor which helps in reminding its content to the audiences (Rousta, Venous, Ebrahimi, 2004). One of considered factors in the debate of evaluating effectiveness of advertisement is the attitude towards advertisement. Attitude towards the advertisement includes public interest or lack of interest in an advertising stimulus in encountering a special topic (Moaen & Minor, 2009). Mental imagery (B one & Elen, 1992), content transparence and emotions called through the advertisement are among variables influencing the attitude towards the advertisement. Some of these factors may lead to strong impacts while some of them may create weak impacts or don't include any impact. Awareness of the impact of each of these factors makes a considerable help in making management decisions (Talis et.al, 2005).

Fotini et.al (2009) provided a model of effective advertisements for advertising websites of big companies. The results of the study showed that measurements, activities, priorities and intentions of people have a considerable impact on evaluation and qualitative factors of relation-oriented marketing have a great impact on advertisements effectiveness indicators. The extracted criteria include:

- Advertisement policies: product information, company information, product general display, consistency with traditional advertisement

- Medium features: message features, information structure, responding time, facilitating research, beauty, interactions, communications in real-time, amusement, information control

- Relation- oriented marketing policy: communications, feedback, supporting customers

Today, organizations for competition in the business world and selling their goods and services require purposeful and efficient marketing activities. One of topics which has a very important role in the process of marketing of an organization and assign a high volume of budget to it is advertisement. Advertisement causes some changes in the knowledge, attitude and behavior level of audiences, so attention to evaluating advertisement effectiveness is a necessity for the organizations that consider advertisement as requirement of continuous life, growth and development. By assessing advertisement effectiveness and investigating consistency with the organization objectives and its expected impacts, some changes may be created in its dimensions, content and methods so that a more effective and useful advertising is performed. Regarding the research literature and internal and external background which has been stated, here a general categorization of the model proposed criteria has been provided in the following table:

Table 1- summation and extracted variables

| scholar | sub standard | criterion |
|---|---|---|
| ) Dargi & Mohammadian (2010), Ashouri (2016), Mansouri Moayed (2005), Aghapour (2016), Mirzadh & Mehrmanesh (2016), Keramforn (2011), Mac Ki- Nesbiat et.al (2011), Rosemary et.al (2012), Wang (2015), Maryion et.al(2017) | awareness of sexuality differences | individual factors |
| Dargi & Mohammadian (2010), Ashouri (2016), Samsam Shariat Keramforn (2011)(2007), Mirzadeh & Mehrmanesh (2016), Mac Ki- Nesbiat et.al (2011), Rosemary et.al (2012), Wang (2015), Maryion et.al(2017) | attention to age requirements | |
| Dargi & Mohammadian (2010), Dehdashti Shahrokh (2009) | various consumers (permanent, ordinary and case consumers) | |

| | | |
|---|---|---|
| Khashizadeh (2010), Danaei & Dasar (2016), Pillay et.al (2012), Marion et.al (2017) | advertisement richness | |
| Danaei & Dasar (2016), Dargi & Mohammadian (2010), Pillay et.al (2012), Olson et.al (2012) | advertisement accuracy | |
| Ashouri (2016), Rabiei (2011), Taeini Harari (2014) | positive mentality of the organization | |
| Dargi & Mohammadian (2010), Shahmohammadi (2011), Olson et.al (2012) | honesty feeling in advertisement | |
| Fakhari et.al (2012) | threatening feeling | |
| Ashourizadeh (2016), Hamidizadeh (2013) | attitude towards brand | |
| madhoushi (2013), Olson et.al (2012) | being convinced | |
| Ahmadi (2013), Mac Ki – Nesbiat et.al (2011), Olson et.al (2012) | convicting audiences | |
| Ashouri (2016), Hosivar (2014), Olson et.al (2012) | mental imagery | |
| Siadat & Sakhi Jourashk (2014) | positive feeling | |
| Navaei et.al (2015), Rozmari et.al (2012) | audience being active | |
| Mosavi Esmaeilzadeh (2015), Olson et.al (2012) | knowledge level | |
| Dargi & Mohammadian (2010), Ashouri (2016), Shafiei Nikabadi & Zarei (2017) | attitude towards advertisement | |
| Dargi & Mohammadian (2010), Hosseinzadeh et.al (2016), Sanei et.al (2016) | familiarity with methods and forms of advertisement | message |
| Ashouri (2016), Olson et.al (2012), Alan (2012), Patric Ro (2014) | attention to reduction of expectation time of the user for observing advertisement | |
| Danaei & Dasar (2016), Forentebury & Mac Gldric (2011) | beautiful and appropriate | |

| | | |
|---|---|---|
| | combination of colors | |
| Aghapuor (2016), Sanei et.al (2016) | site and location of advertisement placement | |
| Hassanzadeh & Danaei (2016), Sanei et.al (2016) | access to sufficient and required information | |
| Dargi & Mohammadian (2010), Darzban (2016) | starting the advertisement with a challenge | |
| Navaei et.al (2015), Salimi Fard & Pourdehghan (2016), Bargar et.al (2016) | starting the advertisement with a statement and explanation | |
| Dargi & Mohammadian (2010), Mirzadeh & Mehrmanesh | using helping words and terms for decision making | |
| Dargi & Mohammadian (2010), Daneshvar et.al (2016), Alan (2012) | using stimuli with diversions more than usual and expected extent | |
| Sanei et.al (2016) | using images of women and children | |
| Sohrabi Ranani (2016), Fortenbury & Mac Goldric (2011) | creativity in the message | |
| Shafei Nikabadi & Zarei (2017), Denis et.al (2012) | the text of advertisement message being a predicate sentence | |

| | | |
|---|---|---|
| Khashizadeh (2010), Dargi & Mohammadian (2010), Sohrabi Ranani (2016), Patric Ro (2014), Abdolsalam et.al (2016) | continuation and repetition of message | |
| Daneshfar & Nabizadeh (2015), Fortenbury & Mac Goldric (2011) | using trustable and famous people | |
| Salimifard & Pourdehghan (2016) | common dialects and terms | |
| Daneshfar & Nabizadeh(2015), Forentnburi & Mac Goldric (2011) | loveliness of the popular person for the observer | |
| Daneshfar & Nabizadeh (2015), Forentnburi & Mac Goldric )(2011 | familiarity of famous person with the observer | |
| Mansouri Moayed (2005), Aghapour (2016) | emotional music | |
| Khonsazadeh (2010), Hosseinzadeh et.al (2016), Aghapour (2016) | using various methods | |
| Siadat & Sakhi Jourashk (2014), Sohrabi Ranani (2016) | high coverage and being universal | |
| Navaei et.al (2016), Mosavi Esmaeilzadeh (2015), Hosseinzadeh et.al (2016), Mirzadeh & Mehrmanesh (2016), Rosemary et.al (2012), Mikit et.al (2012), Spenkar (2013) | media advertisement | |
| Navaei et.al (2015), Mosavi Esmaeilzadeh (2015), Hosseinzadeh et.al (2016), Ashouri (2016), Hassanzadeh & Danaei (2016), Mikit et.al (2012) | multimedia advertisement | |
| Navaei et.al (2015), Mirzadeh & Mehrmanesh (2016) | radio advertisement | |
| Navaei et.al (2015), Mosavi Esmaeilzdeh (2015), Ashouri (2016), Hassanzadeh & Danaei (2016) | cell phone advertisement | |
| Navaei et.al (2015), Hosseinzadeh et.al (2016), Mirzadeh & Mehrmanesh (2016) | creating TV advertisement | |
| Ahmadi (2013), Salimi Fard & Pourdehghan (2016), Hoang et.al (2013) | attraction of rewards | |
| siadat & Sakhi Jourashk (2014), Koumra et.al (2016) | being benevolent | |

| | | |
|---|---|---|
| Darzaian (2016), Zabkouich & Louksten (2011), Abdolsalam et.al (2016) | coordination in speaking and behavior | |
| Ahmadi (2013), Salimi Fard & Pourdehghan(2016), Honari et.al (2015) | reality of messages | |
| Madhoushi (2013), Salimi Fard & Pourdehghan (2016), Houang et.al (2013) | graphic-designing | |
| Shahmohammadi (2011), Madhoushi (2013), Salimi Fard & Pourdehghan(2016) | psychological – geographical division | |
| Mirzadeh & Mehrmanesh (2016), Zabkouich & Lokesten (2011), Menroz et.al (2012), Choubhai et.al (2013), Mokan & Oltz (2014), Ayono et.al (2014) | advertisement on web | |
| Dehdashti Shahrokh (2009), Shahmohammadi (2011), Mirzadeh & Mehrmanesh (2016), Menroz et.al (2012), Choubhai et.al (2013), Mokan & Oltz (2014), Ayono et.al (2014), Henri et.al (2015), Yazghiro- Basteh (2015), Lin (2016) | internet & cell phone advertisement | |
| Shahmohammadi (2011), Sanei et.al (2016), Menroz et.al (2012), Chobai et.al (2013), Ayono et.al (2014), Henri et.al (2015) | selecting advertising media | |
| Shahmohammadi (2011), Sanei et.al (2016), Howang et.al (2013), Henri et.al (2015), Yazghirdo-Yasteh (2015), Martina Santa et.al (2017) | artistic aspect of advertisement | |
| Kashizadeh (2010), Hoshivar(2014), Pelpin et.al (2017) | gender of speaker in the advertisement | |
| Hosseinzadeh et.al (2016), oribeh (2016), Madra et.al (2017) | text repetition | |
| Shahmohammadi (2011), Hassanzadeh & Danaei (2016), Abri & Joumar (2014), Ayouno et.al (2014), Oribeh (2016), Pachman (2016), Bak (2016), Madra et.al (2017) | advertisement | |
| Samsam Shariat (2007), Sohrabi Ranani (2016), Daneshvar et.al (2016), Bak (2016), Kiji-Kali (2017) | factors relating to usage | environment |

| | | |
|---|---|---|
| Fakhari et.al (2012), Mirzadeh & Mehrmanesh (2016), Howang et.al (2013), Bak (2016), kiji-Kali (2017) | factors relating to facilitation | |
| Dargi & Mohammadian (2010), Hamidizadeh (2013), Darzban (2016), Belman et.al (2012) | awareness of different status of the environment status | |
| Dargi & Mohammadian (2010), Mirzadeh & Mehrmanesh (2016) | awareness of access location of objective group to the product | |
| Dargi & Mohammadian (2010), Salimifard & Pourdehghan (2016) | awareness of norms | |
| Dargi & Mohammadian (2010), Hamidizadeh (2013), Darzban (2016), Belman et.al (2012), Koumra et.al (2016) | awareness of religions and religious values | |
| Dargi & Mohammadian (2010), Hassanzadeh & Danaei (2016) | awareness of method of evaluating services in relation to other services | |
| Hassazadeh & Danaei (2016), Radmiker (2015) | attention to security | |
| Samsam Shariat (2007), Mirzadeh & Mehrmanesh (2016), Belman et.al (2012) | economic beliefs | |
| Salimifard & Pourdehghan (2016), Pilay et.al (2012), Howang et.al (2013), Komra et.al (2016) | ethical beliefs | |
| Salimifard & Pourdehghan (2016) | Islamic teaching | |
| Samsam Shariat (2007), Daneshvar et.al (2016), Radmiker (2015) | strategic look to advertisement plan | |
| Dargi & Mohamadian (2010), Ansari & Rasei (2016), Abri & Joumar (2014) | market share | |
| Dargi & Mohamadian (2010), Darzain (2016) | obviating primary needs of audiences | |

| | | |
|---|---|---|
| Salimi Fard & Pourdehghan (2016), Kim et.al (2012), Howang et.al (2013) | reference groups | |
| Sohrabi Ranin (2016), Fontini et.al (2009), Belman et.al (2012) | awareness of services | factors relating to services |
| Salimi Fard & Pourdehghan (2016), Abri & Joumar (2014), Radmiker (2015) | stating potential benefits of purchase ad services consumption | |
| Dargi & Mohamadian (2010), Hosseinzadeh et.al (2016), Fontini et.al (2009) | rate of services complexity | |
| Samsam Shariat (2007), Darzaian (2016), Fontini et.al (2009), Radmiker (2015) | display of expected experience | |
| Khashizadeh (2010), Darzain (2016) | rate of familiarity of consumer with services | |
| Ansari & Rasaei (2016), Belman et.al (2012) | market researches | |
| Hosseinzadeh et.al (2016), Mirzadeh & Mehrmanesh (2016) | record of exploiting services | |
| Dargi & Mohammadian (2010), Ansari & Rasaei (2016), Lousi (2012), Park (2015), Marion et.al (2017) | being unique | |
| Dehdashti Shahrokh (2009), Daneshvar et.al (2016), Radmiker (2015), Jankoski et.al (2016) | price, benefits | |
| Daneshvar et.al (2016), Fontini et.al (2009), Jankoski et.al (2016) | relation with customer | |
| Samsam Shariat (2007), Daneshvar et.al (201), Fontini et.al (2009), Jankoski (2016) | supporting customer | |
| Dehdashti Shahrokh (2009), Ansari & Rasaei (2016), Fontini et.al (2009), Radmiker (2015), Jankoski (2016) | feedback | |

## 2- Research plan

The present study is of descriptive applied type which has been performed with survey method. The statistical population of this study includes all managers and experts of Eghtesad

Novin Bank. The statistical population of the present study contains all people who are exposed to advertisements of Eghtesad Novin Bank across the country. For selecting samples, Cochran formula is used for unlimited societies that based on this formula, the samples number were determined 400 persons for each province. The sample people were selected based on the following sampling method. At first, using stratified sampling method , the samples were randomly selected from five geographical regions of the country (north, south, east, west and center) of nine provinces (Mazandaran, Khorasan Razavi, Zanjan, Chahar Mahal Bakhtiari , Kerman, East Azarbaijan , Kermanshah, Ilam and Tehran).

## 3- Variables and research model

In this study, fuzzy Delphi technique has been used for identifying main indicators. For ranking the final criteria and sub standards, multi-criteria decision making models and AHP model have been used with fuzzy approach. For analysis of obtained data, Excel software and coding in the environment of this software have been used. The results obtained from data analysis have been provided in the present chapter.

### 3-1 Identifying criteria and sub standards of factors effective on advertisement effectiveness

At the first step, the primary indicators were identified and selected. In the first stage, by reviewing the literature and performed studies, a large number of indicators were extracted from the literature. For screening the indicators and identifying the final indicators, fuzzy Delphi approach has been used. The expert views have been collected in the field of rate of significance of main indicators. Fuzzy average and defuzzification output of rates relating to indicators are presented in table 2. The rate of defuzzification greater than 7 is acceptable and every indicator which has the score higher than 7 is confirmed (Wu& Fang, 2011).

Table 2- fuzzy average and fuzzy screening of indicators of factors effective on advertisement effectiveness

| Result | Crisp | mean | U | M | L | |
|---|---|---|---|---|---|---|
| acceptance | 7.15 | (2, 7.97, 9) | 9 | 7.97 | 2 | Q01 |
| acceptance | 7.84 | (6, 8.01, 9) | 9 | 8.01 | 6 | Q02 |

| Result | Crisp | mean | U | M | L | |
|---|---|---|---|---|---|---|
| acceptance | 7.92 | (6, 8.13, 9) | 9 | 8.13 | 6 | Q03 |
| acceptance | 7.39 | (4, 7.83, 9) | 9 | 7.83 | 4 | Q04 |
| acceptance | 7.70 | (6, 7.8, 9) | 9 | 7.80 | 6 | Q05 |
| rejection | 6.68 | (2, 7.28, 9) | 9 | 7.28 | 2 | Q06 |
| acceptance | 7.71 | (6, 7.82, 9) | 9 | 7.82 | 6 | Q07 |
| acceptance | 7.66 | (6, 7.75, 9) | 9 | 7.75 | 6 | Q08 |
| acceptance | 7.02 | (2, 7.78, 9) | 9 | 7.78 | 2 | Q09 |
| acceptance | 7.27 | (2, 8.15, 9) | 9 | 8.15 | 2 | Q10 |
| acceptance | 7.09 | (2, 7.88, 9) | 9 | 7.88 | 2 | Q11 |
| rejection | 6.85 | (2, 7.52, 9) | 9 | 7.52 | 2 | Q12 |
| acceptance | 7.43 | (4, 7.9, 9) | 9 | 7.90 | 4 | Q13 |
| rejection | 6.71 | (3, 7.07, 9) | 9 | 7.07 | 3 | Q14 |
| acceptance | 7.17 | (2, 8.01, 9) | 9 | 8.01 | 2 | Q15 |
| acceptance | 7.13 | (2, 7.95, 9) | 9 | 7.95 | 2 | Q16 |
| rejection | 6.85 | (4, 7.02, 9) | 9 | 7.02 | 4 | Q17 |
| acceptance | 7.43 | (4, 7.89, 9) | 9 | 7.89 | 4 | Q18 |
| acceptance | 7.70 | (6, 7.8, 9) | 9 | 7.80 | 6 | Q19 |
| rejection | 6.80 | (3, 7.21, 9) | 9 | 7.21 | 3 | Q20 |
| acceptance | 7.75 | (6, 7.88, 9) | 9 | 7.88 | 6 | Q21 |
| rejection | 6.62 | (3, 6.93, 9) | 9 | 6.93 | 3 | Q22 |
| acceptance | 7.61 | (6, 7.66, 9) | 9 | 7.66 | 6 | Q23 |
| acceptance | 7.22 | (2, 8.08, 9) | 9 | 8.08 | 2 | Q24 |
| acceptance | 7.04 | (2, 7.81, 9) | 9 | 7.81 | 2 | Q25 |
| acceptance | 7.71 | (6, 7.82, 9) | 9 | 7.82 | 6 | Q26 |
| rejection | 6.65 | (3, 6.97, 9) | 9 | 6.97 | 3 | Q27 |
| acceptance | 7.98 | (6, 8.22, 9) | 9 | 8.22 | 6 | Q28 |
| acceptance | 7.70 | (6, 7.79, 9) | 9 | 7.79 | 6 | Q29 |
| rejection | 6.54 | (3, 6.81, 9) | 9 | 6.81 | 3 | Q30 |
| acceptance | 7.19 | (4, 7.53, 9) | 9 | 7.53 | 4 | Q31 |
| acceptance | 7.79 | (5, 8.18, 9) | 9 | 8.18 | 5 | Q32 |

| Result | Crisp | mean | U | M | L | |
|---|---|---|---|---|---|---|
| acceptance | 7.53 | (4, 8.05, 9) | 9 | 8.05 | 4 | Q33 |
| acceptance | 7.05 | (2, 7.82, 9) | 9 | 7.82 | 2 | Q34 |
| acceptance | 7.43 | (4, 7.89, 9) | 9 | 7.89 | 4 | Q35 |
| rejection | 6.92 | (3, 7.38, 9) | 9 | 7.38 | 3 | Q36 |
| acceptance | 7.02 | (2, 7.78, 9) | 9 | 7.78 | 2 | Q37 |
| rejection | 6.91 | (3, 7.37, 9) | 9 | 7.37 | 3 | Q38 |
| rejection | 6.96 | (2, 7.68, 9) | 9 | 7.68 | 2 | Q39 |
| rejection | 6.99 | (2, 7.74, 9) | 9 | 7.74 | 2 | Q40 |
| acceptance | 7.42 | (4, 7.88, 9) | 9 | 7.88 | 4 | Q41 |
| acceptance | 7.10 | (2, 7.9, 9) | 9 | 7.90 | 2 | Q42 |
| rejection | 6.59 | (2, 7.13, 9) | 9 | 7.13 | 2 | Q43 |
| rejection | 6.55 | (2, 7.08, 9) | 9 | 7.08 | 2 | Q44 |
| acceptance | 7.77 | (6, 7.9, 9) | 9 | 7.90 | 6 | Q45 |
| rejection | 6.91 | (2, 7.62, 9) | 9 | 7.62 | 2 | Q46 |
| acceptance | 7.09 | (2, 7.88, 9) | 9 | 7.88 | 2 | Q47 |
| acceptance | 7.68 | (6, 7.78, 9) | 9 | 7.78 | 6 | Q48 |
| acceptance | 8.19 | (6, 8.53, 9) | 9 | 8.53 | 6 | Q49 |
| rejection | 6.88 | (2, 7.57, 9) | 9 | 7.57 | 2 | Q50 |
| acceptance | 7.75 | (5, 8.12, 9) | 9 | 8.12 | 5 | Q51 |
| acceptance | 7.03 | (2, 7.8, 9) | 9 | 7.80 | 2 | Q52 |
| rejection | 6.96 | (4, 7.19, 9) | 9 | 7.19 | 4 | Q53 |
| acceptance | 7.04 | (2, 7.8, 9) | 9 | 7.80 | 2 | Q54 |
| rejection | 6.96 | (3, 7.45, 9) | 9 | 7.45 | 3 | Q55 |
| acceptance | 7.02 | (2, 7.78, 9) | 9 | 7.78 | 2 | Q56 |
| acceptance | 7.92 | (5, 8.38, 9) | 9 | 8.38 | 5 | Q57 |
| acceptance | 7.00 | (2, 7.76, 9) | 9 | 7.76 | 2 | Q58 |
| rejection | 6.90 | (1, 7.86, 9) | 9 | 7.86 | 1 | Q59 |
| acceptance | 7.10 | (4, 7.4, 9) | 9 | 7.40 | 4 | Q60 |
| acceptance | 7.86 | (6, 8.04, 9) | 9 | 8.04 | 6 | Q61 |
| rejection | 6.97 | (2, 7.71, 9) | 9 | 7.71 | 2 | Q62 |

| Result | Crisp | mean | U | M | L | |
|---|---|---|---|---|---|---|
| acceptance | 7.02 | (2, 7.79, 9) | 9 | 7.79 | 2 | Q63 |
| acceptance | 7.73 | (6, 7.84, 9) | 9 | 7.84 | 6 | Q64 |
| acceptance | 7.01 | (2, 7.76, 9) | 9 | 7.76 | 2 | Q65 |
| rejection | 6.65 | (3, 6.98, 9) | 9 | 6.98 | 3 | Q66 |
| acceptance | 7.01 | (2, 7.77, 9) | 9 | 7.77 | 2 | Q67 |
| acceptance | 7.90 | (6, 8.11, 9) | 9 | 8.11 | 6 | Q68 |
| acceptance | 7.55 | (5, 7.83, 9) | 9 | 7.83 | 5 | Q69 |
| acceptance | 8.02 | (6, 8.28, 9) | 9 | 8.28 | 6 | Q70 |
| acceptance | 7.95 | (6, 8.17, 9) | 9 | 8.17 | 6 | Q71 |
| acceptance | 7.03 | (2, 7.79, 9) | 9 | 7.79 | 2 | Q72 |
| acceptance | 7.65 | (5, 7.98, 9) | 9 | 7.98 | 5 | Q73 |
| acceptance | 7.00 | (1, 8, 9) | 9 | 8.00 | 1 | Q74 |
| acceptance | 7.04 | (2, 7.8, 9) | 9 | 7.80 | 2 | Q75 |
| acceptance | 7.03 | (2, 7.8, 9) | 9 | 7.80 | 2 | Q76 |
| acceptance | 7.30 | (4, 7.7, 9) | 9 | 7.70 | 4 | Q77 |
| rejection | 6.88 | (2, 7.57, 9) | 9 | 7.57 | 2 | Q78 |
| acceptance | 7.63 | (4, 8.19, 9) | 9 | 8.19 | 4 | Q79 |
| acceptance | 7.65 | (5, 7.98, 9) | 9 | 7.98 | 5 | Q80 |
| rejection | 6.59 | (3, 6.88, 9) | 9 | 6.88 | 3 | Q81 |
| acceptance | 7.09 | (2, 7.88, 9) | 9 | 7.88 | 2 | Q82 |
| rejection | 6.83 | (2, 7.5, 9) | 9 | 7.50 | 2 | Q83 |
| acceptance | 7.12 | (2, 7.93, 9) | 9 | 7.93 | 2 | Q84 |

From the 84 indicators which were assessed, 24 indicators which have acquired the average score less than 7 are omitted. So, 60 indicators are finally confirmed (Adelazaret.al, 1995).

The remained indicators will be finally examined in the format of a 60-question questionnaire with analytical hierarchy process.

3-2 determining the priority of factors effective on effectiveness of advertisement using fuzzy AHP technique

For determining the priority of major criteria, fuzzy analytical hierarchy (FAHP) has been used. The analysis process is as follows.

1- Dual comparison of major criteria based on objective and determining the weight of major criteria

2- Dual comparison of sub standards of each criterion and determining sub standards of each cluster

3- Multiplication of the weight of sub standards in the weight of respective criterion and determining the final weight of sub standards

For dual comparison of elements, 9-degree Saaty scale has been used. 9-degree Saaty scale has been provided by Thomas Saaty , codifier of analytical hierarchy theory. Also, in this study fuzzy approach has been used for making the values quantitative. So, Saaty fuzzy range has been used.

Table 3- fuzzy range equal to 9-degree Saaty scale in AHP technique. Habibi et.al , 2014, 77

| reversed fuzzy equivalent | fuzzy equivalent | the word term of comparison status of i to j |
|---|---|---|
| (1,1,1) | (1, 1, 1 ) | Preferred Equally |
| $\left(\frac{1}{3},\frac{1}{2},1\right)$ | (1, 2, 3 ) | intermediate |
| $\left(\frac{1}{4},\frac{1}{3},\frac{1}{2}\right)$ | (2, 3, 4 ) | Preferred moderately |
| $\left(\frac{1}{5},\frac{1}{4},\frac{1}{3}\right)$ | (3, 4, 5 ) | intermediate |
| $\left(\frac{1}{6},\frac{1}{5},\frac{1}{4}\right)$ | (4, 5, 6 ) | Preferred Strongly |
| $\left(\frac{1}{7},\frac{1}{6},\frac{1}{5}\right)$ | (5, 6, 7 ) | intermediate |
| $\left(\frac{1}{8},\frac{1}{7},\frac{1}{6}\right)$ | (6, 7, 8 ) | very strongly Preferred |
| $\left(\frac{1}{9},\frac{1}{8},\frac{1}{7}\right)$ | (7, 8, 9 ) | intermediate |
| $\left(\frac{1}{9},\frac{1}{9},\frac{1}{9}\right)$ | (9, 9, 9 ) | Extremely Preferred |

( source in papersL Khorshid & Zabihi, Lee my H.I; Wen-Chin Chen, Ching-Jan, Chang, 2008, 101)

The pattern of criteria and sub standards has been drawn in the figure using AHP technique. Also, the criteria and sub standards of the study have been named with numeric index so that the research flow is easily detectable.

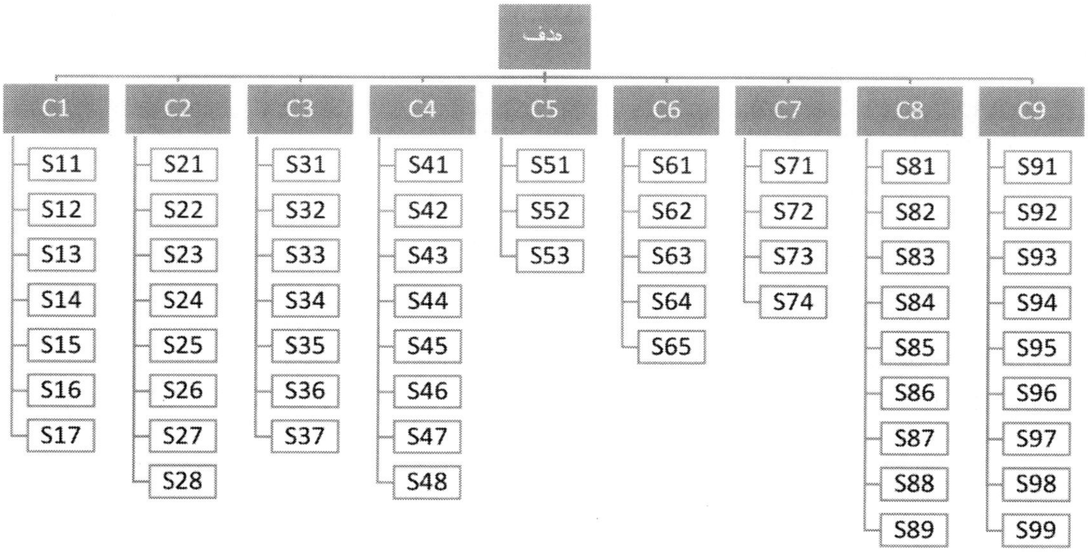


Figure 1- hierarchical model of criteria and sub standards of factor effective on advertisement effectiveness

**1- Determining the priority of major criteria based on objective**

In the first step, the major criteria based on the objective have been compared dually. Dual comparison is very simple and all elements of each cluster should be compared two by two. Therefore, if there is n element in a cluster, $\frac{n(n-1)}{2}$ comparisons will be performed. Since there are four criteria, so the number of performed comparisons is equal to:

$$\frac{n(n-1)}{2} = \frac{9(9-1)}{2} = 36$$

So, 36 dual comparisons have been performed from group perspective of experts. The experts view has been made quantitative using fuzzy scale. Collecting experts view has been collected with 9-degree Saaty range. So, 6 dual scales of four major four criteria have been performed based on the experts view. Using geometric average the experts views have been aggregated. Fuzzy aggregation will be calculated as follows:

Equation 1

$$F_{AGR} = (\prod(l), \prod(m), \prod(u))$$

After matrix formation of the obtained dual comparisons, the special vector has been calculated. At first, the fuzzy extension of each row is calculated. Every entry of dual comparison matrix $\tilde{X}$

Is displayed as $\tilde{x}_{ij}$ .

The fuzzy extension of every row is also displayed with symbol $\tilde{S}_i$. Therefore, the fuzzy extension of each row is calculated as follows:

Equation 2

$$\tilde{S}_i = \sum_{j=1}^{n} x_{ij}$$

Then, the fuzzy extension of every row elements will be as follows:

Fuzzy extension of row 1 (1,1,1) ⊕ (1,27,1,47,1,75) ⊕ (0.47, 0.55, 0.66) ⊕ (1.52, 1.81, 2.14) ⊕ (0.93, 1.14, 1.39) ⊕ (1.26, 1.61, 2.04) ⊕(1.25, 1.51, 1.84) ⊕(0.78, 0.97, 1.24) ⊕ (0.59, 0.75, 0.96) ⊕ (9.08, 10.8, 13.02)

Fuzzy extension of row 2 (0.57, 0.68, 0.79)⊕(1, 1, 1)⊕(0.58, 0.7, 0.89)⊕(1.84, 2.28, 2.75)⊕(0.68, 0.82, 1.02)⊕(1.42, 1.81, 2.34)⊕(1.64, 2.04, 2.53)⊕(0.71, 0.88, 1.12)⊕(0.73, fuzzy extension of row 3 (1.52, 1.82, 2.11)⊕(1.13, )0.9, 1.1)⊕ = (9.17, 11.11, 13.55 1.42, 1.72)⊕(1, 1, 1)⊕(1.89, 2.34, 2.88)⊕(0.7, 0.86, 1.09)⊕(1.45, 1.78, 2.22)⊕(1.55, 1.88, 2.29)⊕(0.57, 0.71, 0.93)⊕(0.67, 0.8, 0.96)⊕ = (10.47, 12.61, 15.2)

Fuzzy extension of row 4 (0.47, 0.55, 0.66)⊕(0.36, 0.44, 0.54)⊕(0.35, 0.43, 0.53)⊕(1, 1, 1)⊕(0.97, 1.21, 1.51)⊕(0.96, 1.33, 1.8)⊕(2.14, 2.7, 3.37)⊕(0.94, 1.21, 1.56)⊕(1.3, 1.69, 2.16)⊕ = (8.49, 10.57, 13.14)

Fuzzy extension of row 5 (0.72, 0.88, 1.07) ⊕ (0.98, 1.22, 1.48)⊕(0.92, 1.17, 1.43)⊕(0.66, 0.82, 1.03)⊕(1, 1, 1)⊕(0.67, 0.83, 1.04)⊕(0.23, 0.29, 0.36)⊕(0.28, 0.32, 0.39)⊕(0.3, 0.38, 0.48)⊕ = (5.76, 6.91, 8.29)

Fuzzy extension of row 6 (0.49, 0.62, 0.79)⊕(0.43, 0.55, 0.7)⊕(0.45, 0.56, 0.69)⊕(0.55, 0.75, 1.04)⊕(0.96, 1.2, 1.48)⊕(1, 1, 1)⊕(0.34, 0.41, 0.51)⊕(0.27, 0.34, 0.44)⊕(0.24, 0.31, 0.41)⊕ = (4.74, 5.75, 7.08)

Fuzzy extension of row 7 (0.54, 0.66, 0.8)⊕(0.4, 0.49, 0.61)⊕(0.44, 0.53, 0.64)⊕(0.3, 0.37, 0.47)⊕(2.81, 3.5, 4.3)⊕(1.95, 2.43, 2.96)⊕(1, 1, 1)⊕(0.34, 0.42, 0.54)⊕(0.29, 0.36, 0.47)⊕ = (8.06, 9.76, 11.78)

Fuzzy extension of row 8 (0.8, 1.03, 1.28)⊕(0.89, 1.14, 1.41)⊕(1.08, 1.41, 1.76)⊕(0.64, 0.82, 1.06)⊕(2.55, 3.1, 3.63)⊕(2.25, 2.97, 3.72)⊕(1.86, 2.39, 2.94)⊕(1, 1, 1)⊕(0.58, 0.69, 0.83)⊕ = (11.66, 14.56, 17.64)

Fuzzy extension of row 9 (1.04, 1.33, 1.69)⊕(0.91, 1.11, 1.37)⊕(1.04, 1.25, 1.5)⊕(0.46, 0.59, 0.77)⊕(2.08, 2.63, 3.3)⊕(2.42, 3.23, 4.1)⊕(2.15, 2.74, 3.43)⊕(1.2, 1.46, 1.72)⊕(1, 1, )1)⊕ = (12.29, 15.35, 18.88

Then, the fuzzy sum of total priorities column elements is calculated

Equation 3

$$\sum \tilde{S}_i = \sum_{i=1}^{n} \sum_{j=1}^{n} x_{ij}$$

The sum of elements of the priorities column of major criteria will be as follows:

$$\sum_{i=1}^{4} \sum_{j=1}^{4} M_g^j = (79.72, 97.41, 118.59)$$

For normalization of each criterion priorities, that criterion total rates sum should be divided into sum of all priorities (column elements), since the rates are fuzzy, so fuzzy sum of every row is multiplied in summation reverse. The summation reverse should be calculated.

Equation 4

$$if\ \tilde{F} = (l, m, u)\ then\ \tilde{F}^{-1} = \left(\frac{1}{u}, \frac{1}{m}, \frac{1}{l}\right)$$

Then, based on the equation 4, we will have

$$\left(\sum_{i=1}^{n} \sum_{j=1}^{n} M_g^j\right)^{-1} = (0.01, 0.01, 0.01)$$

$$P_k = \sum_{i=1}^{n} M * \left(\sum_{i=1}^{n} \sum_{j=1}^{n} M_g^j\right)^{-1}$$

So, the results obtained from normalization of obtained rates will be as follows:

$\tilde{W}_{C1} =$ (0.08, 0.11, 0.16)
$\tilde{W}_{C2} =$ (0.08, 0.11, 0.17)
$\tilde{W}_{C3} =$ (0.09, 0.13, 0.19)
$\tilde{W}_{C4} =$ (0.07, 0.11, 0.16)
$\tilde{W}_{C5} =$ (0.05, 0.07, 0.1)

$\widetilde{W}_{C6} =$ (0.04, 0.06, 0.09)

$\widetilde{W}_{C7} =$ (0.07, 0.1, 0.15)

$\widetilde{W}_{C8} =$ (0.1, 0.15, 0.22)

$\widetilde{W}_{C9} =$ (0.1, 0.16, 0.24)

Each of the obtained rates is fuzzy and normalized weight relating to major criteria. There are various methods for defuzzifcation of the obtained rates. In this study, Bojadziev proposed method has been used for defuzzification.

Table 4- defuzzification and computation of final and normal weight of criteria of factors effective on advertisement effectiveness

| Normal | Deffuzy | X3max | X2max | X1max | Crisp |
|---|---|---|---|---|---|
| 0/111 | 0/117 | 0/113 | 0/115 | 0/117 | C1 |
| 0/114 | 0/120 | 0/115 | 0/118 | 0/120 | C2 |
| 0/103 | 0/137 | 0/133 | 0/135 | 0/137 | C3 |
| 0/108 | 0/113 | 0/112 | 0/113 | 0/113 | C4 |
| 0/070 | 0/073 | 0/072 | 0/073 | 0/073 | C5 |
| 0/060 | 0/063 | 0/062 | 0/063 | 0/063 | C6 |
| 0.101 | 0/107 | 0/103 | 0/105 | 0/107 | C7 |
| 0.149 | 0/157 | 0/153 | 0/155 | 0/157 | C8 |
| 0.158 | 0/167 | 0/163 | 0/165 | 0/167 | C9 |

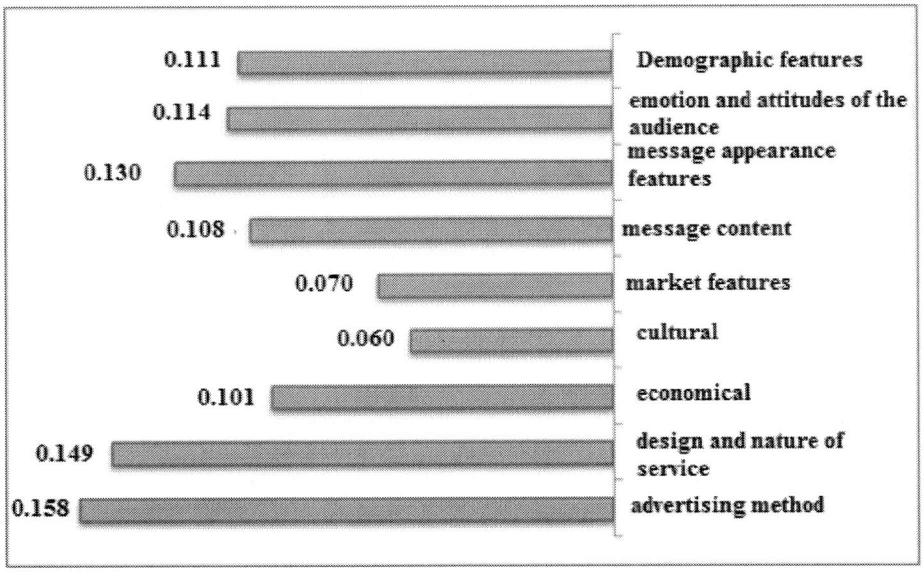


Figure 2- graphical display of significance of criteria of factors effective on advertisement effectiveness

Based on the obtained special vector:

- Demographic features with weight 0.111 have been located in rank five

- Individual feelings and attitude of the addressee with weight 0.114 has been located in rank four.

- The message surface features with weight 0.13 has been located in rank three.

- The message content with weight 0.108 has been located in rank six

- The market feature with weight 0.07 has been located in rank eight.

- Cultural with weight 0.06 has been located in rank nine.

- Economic with weight 0.10 has been located in rank seven.

- Plan and nature of services with weight 0.149 has been located in rank second.

- Advertisement methods with weight 0.158 has been located in rank first.

The inconsistency rate of performed comparisons has been obtained 0.030 which is smaller than 0.1 and so the performed comparisons is sufficed.

**2- Comparison and determining priority of sub standards**

At first, the experts' views have been collected in the dual comparison of sub standards. Then, the performed calculations for fuzzification of the experts views average have been provided for determining the priority of sub standards.

Priority of sub standards of demographic features

The audience job with weight 0.151 has been located in the third rank.

The audience income with weight 0.093 has been located in the seventh rank.

The audience religion with weight 0.16 has been located in the second rank.

Social class with weight 0.141 has been located in the fifth rank.

Awareness of gender differences with weight 0.173 has been located in the first rank.

The individual knowledge level with weight 0.135 has been located in the sixth rank.

Priority of sub standards of the audience individual feeling and attitude

Positive mentality with weight 0.124 has been located in the fifth rank.

Feeling of honesty in advertisements with weight 0.089 has been located in the eighth rank.

Threatening feeling with weight 0.14 has been located in the third rank.

Mental image with weight 0.102 has been located in the seventh rank.

Attitude towards brand with weight 0.121 has been located in the sixth rank.

Convincing audiences with weight 0.149 has been located in the first rank.

Attitude towards advertisement with weight 0.143 has been located in the second rank.

The individual perception of accuracy of advertisement with weight 0.133 has been located in the fourth rank.

Priority of sub standards of the message surface features

Then, based on the results of performed dual comparisons:

Beautiful and appropriate combination of colors with weight 0.143 has been located in the fourth rank.

Attention to reduction of the user waiting time with weight 0.159 has been located in the second rank.

Graphics- design with weight 0.131 has been located in the sixth rank.

Creativity in the message with weight 0.137 has been located in the fifth rank.

Emotional music with weight 0.099 has been located in the seventh rank.

Text repetition with weight 0.153 has been located in the third rank.

Access to required and sufficient information with weight 0.178 has been located in the first rank.

Determining the priority of sub standards of the message content

Benevolent with weight 0.093 has been located in the eighth rank.

Coordination in speech and behavior with weight 0.125 has been located in the fourth rank.

Reality of the messages with weight 0.119 has been located in the sixth rank.

Using women and children images with weight 0.131 has been located in the third rank.

Gender of speaker in the advertisement with weight 0.106 has been located in the seventh rank.

Using trustful and famous people with weight 0.16 has been located in the first rank.

Loveliness of famous people for the observer with weight 0.122 has been located in the fifth rank.

Familiarity of the famous person with the observer with weight 0.144 has been located in the second rank.

Priority of sub standards of the market features

Then, based on the results of the performed dual comparisons:

Factors relating to its application in the market with weight 0.326 have been located in the second rank.

Awareness of different aspects of status and environment with weight 0.283 has been located in the third rank.

Awareness of evaluation method of service receiver with weight 0.391 has been located in the first rank.

Priority of cultural sub standards

Then, based on the results of performed dual comparisons:

Awareness of norms with weight 0.199 has been located in the third rank.

Awareness of religion and religious values with weight 0.238 has been located in the first rank.

Ethical beliefs with weight 0.18 have been located in the fourth rank.

Obviating primary needs of the audiences with weight 0.164 has been located in the fifth rank.

Reference groups with weight 0.219 have been located in the second rank.

Determining the priority of economic sub standards

Then, based on the results of the performed dual comparison:

Economic beliefs with weight 0.261 have been located in the second rank.

Strategic look at an advertising program with weight 0.215 has been located in the fourth rank.

Market share with weight 0.274 has been located in the first rank.

Attention to economic security with weight 0.251 has been located in the third rank.

Priority of the plan and services nature sub standard

Awareness of services with weight 0.104 has been located in seventh rank.

Expressing potential benefits of purchase and services consumption with weight 0.117 has been located in the third rank.

Services complexity rate with weight 0.089 has been located in the ninth rank.

Uniqueness with weight 0.108 has been located in the sixth rank.

Rate of familiarity of the services consumer with weight 0.12 has been located in the second rank.

Price and allowances with weight 0.114 has been located in the fourth rank.

Relation with the customer with weight 0.111 has been located in the fifth rank.

Displaying the expected experience with weight 0.098 has been located in the eighth rank.

Supporting customers with weight 0.139 has been located in the first rank.

Priority of sub standards of the advertising methods

Beginning the advertisement with a challenge with weight 0.079 has been located in the ninth rank.

Beginning an advertisement with a statement and explanation with weight 0.107 has been located in the fifth rank.

Using stimuli with diversions more than usual and expected with weight 0.097 has been located in the seventh rank.

The advertisement message text being predicating with weight 0.126 has been located in the third rank.

Using diverse methods with weight 0.088 has been located in the eighth rank.

Media advertisement with weight 0.113 has been located in the fourth rank.

Dimensions of TV advertisement with weight 0.151 have been located in the first rank.

Psychological, geographical division with weight 0.107 has been located in the fifth rank.

Advertising media selection with weight 0.132 has been located in the second rank.

Determining the final priority of major criteria with FAHP technique

For determining final priority of major criteria using FAHP technique, the weights relating to major criteria (W1) and weight of indicators based on each criteria (W2) are needed. The results of comparison of the research sub criteria relating to them forms matrix W2. For determining the final priority of indicators with AHP technique, it is sufficient that the weight of indicators based on every criterion (W2) is multiplied in the major criteria (W1). Any of these matrices has been calculated in the previous steps. Based on the obtained output, it is obvious that the market share item, one of economic factors with weight 0. 277 is the most important indicator in the existing indicators. Item of awareness of evaluation method of service receiver in relation to other services with weight of 0.0273 has the second priority. Item of economic believes, one of economic factors, has the third priority.

So, the research model is as follows:

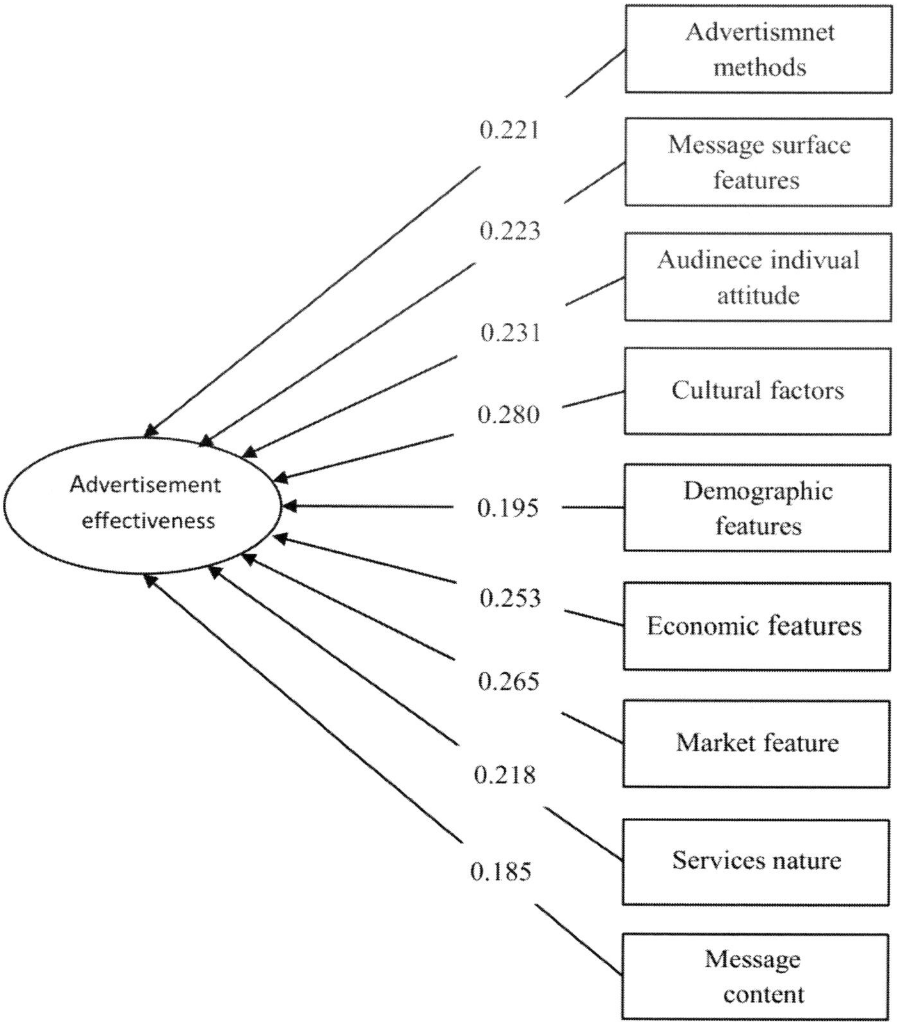


**4- Conclusion:**

In this study, identification and ranking of factors effective on designing the native model of advertisement in Eghtesad Novin bank has been addressed. In this first step, the primary indicators were identified and selected. In the first stage, by reviewing the literature and performed studies, a great number of indicators were extracted from the literature. For screening the indicators and identifying final indicators, fuzzy Delphi approach has been used. The experts' views have been collected in the field of significant rate of major indicators. From 84 indicators which were assessed, 24 indicators which have acquired score less than 7 are omitted. So, 60 indicators were finally confirmed. Then, regarding the performed calculations, the final weight of any of the model indicators have been calculated with fuzzy AHP. Based on the obtained output, it is obvious that the market share of economic factors with weight of 0.027 is the most important indicator in the existing indicators. The item of awareness of evaluating services in relation to other services is one of market features factors with weight 0.0273 which has second priority. The item of economic beliefs with weight 0.0264 is in the third priority.

Today, considering extension of competition in the international level and formation of world competition, all countries of the world may be successful in this arena to coordinate them with this trend and create more value for their customers by services and products they provide. In the today modern world, almost all people are exposed to advertisements and other communicational activities. Effectiveness of advertisement is one of important issues in the advertisement domain. In this regard, Mansouri Moyed (2005) showed that attention attraction, commercial advertisements of Maskan Bank are effective but in respect of other variables, that is, creating interest, stimulation of tendency and measurement to purchase, commercial advertisements of Maskan Bank are not effective. Samsam Shariat (2007) showed that TV channels and cartoon animations , attractive package, colors relating to the product and innovation, noticing safety feeling and associated symbols, informing of quality, background and price, dialects and common phrases, diversity in the message content, in TV, broadcasting from 18-23 and before beginning programs (in media) have been effective factors in effectiveness of commercial effectiveness from people viewpoint. Mosavi Esmaeilzadeh (2015) showed that oral advertisements (in 4 dimensions) and after that TV advertisement (just in 2 dimensions) are effective. Also, radio, internet and billboard advertisement are not effective. And also Friedman test was used for ranking advertisement tools. Salimi Fard & Pourdehghan (2016) showed that among factors effective on attitude to cell phone advertisement, components of general attitude towards advertisement, knowledge level and value have respectively the greatest significance and components of time of presentation and reference groups with the least significance in effecting the individuals attitude towards cell phone advertisement. Ansari & Rasaei (2016) showed that the creativity of advertisement message , market research, competition, market share, uniqueness and relation with customer are effective on the effectiveness of brand advertisement. Louis (2012) showed that advertising in the urban public spaces will have a great impact in the consumers' mentality. Since some advertisements are transitory and are often displayed in a short period of time and just in several spaces, but advertisement in public spaces exist everywhere and is similar to advertising through print and media, but since they have lower cost for production and distribution, have lower cost in this respect. Zabckowich & Loukston (2011) showed that in necktie advertisements use more of sensory requests and information signs. Findings showed that in western and European countries, people establish more contact due to globalization increase. They declared that content of printed advertisements, includes visual and text advertisements which are effective in establishing relation with many audiences. Also, ethnic identity as an important cultural variable increases attention to consumer purchase behavior. Besides that, they found out that the image of people, models, ceremonies and managers constitute an important part of images and background image plays an important role in effectiveness of visual factors in advertisement. Mac Ki- Nesbiat et.al (2011) showed that younger adults remind emotional messages specially negative ones better than others and older adults more prefer positive and logical messages to negative and emotional ones, but attitude towards advertisement among younger adults is not different. Also, they found that older adults have the most positive attitude towards advertisements.

Perzouti (2015) showed that advertising models are different in various ages and each age is consistent with a special model and using various and consistent models with people conditions may make the advertisement more effective. Abri & Joumar (2014) showed that advertising messages are effective on the consumer behavior and the advertiser may use various media for communicating consumer and media schedule is effective on the advertisement success. Attention to environmental factors where the customer lives are important factors in the advertisement success. Madra et.al (2017) showed that the advertisement text, text repetition impact the advertisement effectiveness. Maryion et.al (2017) showed that simultaneous use of several media like TV and internet may increase effectiveness of the advertisement and considering two variables of gender and multimedia create a better individual attitude towards the message effectiveness.

In ranking the identified factors, advertisement methods have assigned the first rank among factors. It is suggested that in designing advertisement and selecting advertising instruments, the dimensions of TV advertising, selecting median and predicative message text should be noticed. Using advertising instrument like internet advertisement, adverting on public vehicles like buses and metro due to great number of people who use them can increase stability of the advertisements in the customers mind.

In codifying and designing bank advertisements, the design and services nature should be noticed. It is proposed that supporting customer, familiarity of consumer with services are considered and during the advertisement process, it is required that the potential benefits of purchase and services consumption are expressed.

In designing the advertisement and content of bank advertising message, it is required to use famous and trustful people for introducing the services. Also, children as future customers and audiences should be noticed since they are future potentials. Total advertisements provided by the bank through various media should move in a unit line so that the audience can establish a logical relation between the observed advertisements in various media that is total advertising efforts lead to synergy of advertisement effectiveness. This issue won't be acquired unless through a media planning and efficient advertisement.

In advertisements, the message surface features should be noticed. It is required that in advertisement, access to required and enough information for the customers, attention to reduction of expectation time of user for observing the advertisement and suitable repetition of the text are considered. Providing large information in an advertisement which has a short time and providing information that require special mathematical analysis should be avoided.

**References:**

*Hosseinabadi, Maryam, Mahmoud Nematian, Mohammad Besharati, investigating the relation between mixed elements of marketing and customers satisfaction with Tehran*

*Sarmayeh bank branches, management and entrepreneurship studies, fall 2016, 2nd period, No. 3 (page 162-169)*

*Dargi Parviz, 2008, subjects and topics of marketing management with Iran market attitude, Tehran, Rasa publication*

*Sarmad, Zohreh, Bazargan Abbas, Hejazi, Elaheh, 2014, research methods in behavioral sciences, Tehran, Agah. 27-13*

*Salimi Fard, Khodakaram, Pourdehghan, Adel, 2016, identifying and ranking factors effective on attitude towards cell phone advertisement using AHP, Brand management, period 3, no. 6, summer 2016, page 133-160*

*Seyed Akbar Manjgani, Vahideh Tabasi Lotfabadi, 2015, factors effective on effectiveness of advertisement on financial stoop of legal clients of Mashhad City with AHP approach, finance journal, fall 2015, No. 75 (page 121-146)*

*Sanayei, Ali, Majid Mohammad Shafiei, Maedeh Amini Valashani, 2016, effectiveness of bank advertisement in customer attraction, commercial management searches, spring and summer 2016, No. 15 (page 185-210)*

*Mahmoudian, Mahmoud, Rabiei, Ali, Jamili, Bita. 2011, evaluation of effectiveness of Parsian bank advertisement and identifying the most important factors in increasing its effectiveness in Tehran city, modern marketing studies, summer 2011, No. 2 (page 17-40).*

*Manjagani, Seyed Akbar, Tabasi Lotfabadi, Vahideh, 2015, factors effective on advertisement effectiveness on financial stoop of legal clinets of Mashhad city with AHP approach, finance journal, fall 2015, No. 75, 1-121*

*Venous, Davar, Safaeian, Mitra, (2005), marketing bank services in Iran and bank services features, Tehran, Negah Danesh, 4th edition*

*Angulo, N. (2016). 10 digital marketing trends to watch in 2016 and beyond. Retrieved from: http://www.marketingdive.com/news/10-digital-marketing-trends-to-watch-in-2016-./and-beyond/411457*

*Bell Raoul, Buchner, Axel.(2018). Positive Effects of Disruptive Advertising on Consumer Preferences, Journal of Interactive Marketing 41 (2018) 1–13*

*Corvi. Elisabetta.(2010).The effectiveness of advertising: a literature review, 10th Global Conference on Business & Economics, ISBN : 978-0-9830452-1-2*

*Cramphorn Michael F.,(2011), Forum Gender effects in advertising, International Journal of .Market Research Society Vol. 53 Issue 2*

*Cronin, J. J. & Fox , G. L. (2010). "The implications of third party customer complaining for .advertising efforts". Journal of advertising. 39( 2): 21-23*

*Decrop, A. (2007). The influence of message format on the effectiveness of print advertisements for tourism destinations. International Journal of Advertising, 26(4), 505–.525*

*Denis .A. Shakhov and ,Anna A. Panasenko.(2012).Evaluating Effectiveness of Bank Advertising in the Internet: Theory and Practice, World Applied Sciences Journal 18 (Special Issue of Economics): 83-90, 2012 ISSN 1818-4952*

*Denis 1,2 A. Shakhov and 3Anna A. Panasenko.(2012).Evaluating Effectiveness of Bank Advertising in the Internet: Theory and Practice, World Applied Sciences Journal 18 (Special Issue of Economics): 83-90, 2012*

*Durdana Ozretic-Dosen Ines Zizak , (2015),"Measuring the quality of banking services targeting student population", EuroMed Journal of Business, Vol. 10 Iss 1 pp. 98 - 117*

*Eckler, P., and Bolls, P. (2011). Spreading the Virus: Emotional Tone of Viral Advertising and Its Effect on Forwarding Intention and Attitudes. Journal of Interactive Advertising, .11(2): 1-11*

*Eric W.T. Ngai Ka-leung Karen Moon S.S. Lam Eric S. K. Chin Spencer S.C. Tao , (2015),"Social media models, technologies, and applications", Industrial Management & Data Systems, Vol. 115 Iss 5 pp. 769 – 802*

*Fotini Patsioura Maro Vlachopoulou Vicky Manthou, (2009),"A new advertising effectiveness model for corporate advertising web sites", Benchmarking: An International Journal, Vol. 16 Iss 3 pp. 372 - 386*

*Frederic Crane, , G. (2010). «Marketing FO Entrepreneurs: ConceptandApplication for new .Venture. New York: SAGE Pablication, Inc*

*Frolova. Svetlana.(2014).THE ROLE OF ADVERTISING IN PROMOTING A PRODUCT, Thesis CENTRIA UNIVERSITY OF APPLIED SCIENCES ,Degree Programme in Industrial Management May 2014, p"1-61*

*Henry Boateng Abednego Feehi Okoe , (2015),"Consumers' attitude towards social media advertising and their behavioural response", Journal of Research in Interactive Marketing, Vol. 9 Iss 4 pp. 299 -312*

*Henry Boateng Abednego Feehi Okoe , (2015),"Consumers' attitude towards social media advertising and their behavioural response", Journal of Research in Interactive Marketing, Vol. 9 Iss 4 pp. 299 -312*

*Huang. Jinsong, Song Sub, , Liuning Zhouc, , , Xi Liu .(2013).Attitude Toward the Viral Ad: Expanding Traditional Advertising Models to Interactive Advertising, Journal of Interactive Marketing, Volume 27, Issue 1, February 2013, Pages 36–46*

*Idris, I., M.S.A. Yajid and A. Khatibi, 2009. Personal and Psychological Factors-Does it .Impact the Choice of Advertising Medium? Journal of Social Sciences, 5(2): 104-111*

*Ioannou Myria, Nikolaos Boukas, Evi Skoufari .(2014).Examining the role of advertising on the behaviour of co-operative bank consumers, Journal of Co-operative Organization and Management, Volume 2, Issue 1, June 2014, Pages 24-33*

*Jankowski. Jarosław, Przemysław Kazienko, Jarosław Wątróbski, Anna Lewandowska, Magdalena Ziolo.(2016).Fuzzy multi-objective modeling of effectiveness and user experience in online advertising, Expert Systems with Applications, Volume 65, 15 December 2016, Pages 315-331*

*Kim, C., Park, S., Kwon, K., & Chang, W. (2012). An empirical test to measure the effectiveness of online advertising in online marketplaces using a hierarchical Bayes model. .Expert Systems with Applications (39), 117–128*

*Kim, ji-Hern., Hyun, yong j., (2010). " A model to investigate the influence of marketing – mix efforts and corporate image on brand equity in the IT software sector " . industrial . Marketing Management. No 40,2011, pp 424-438, Emerald Group Publishing Limited*

*Kim, S., & Jun, J. (2016). The impact of event advertising on attitudes and visit intentions. .Journal of Hospitality and Tourism Management, 29, 1-8*

*Kirby, J. and Marsden .(2006). Connected marketing : the viral, buzz and word of mouth .revolution, Oxford : Butterworth-Heinemann, 2006*

*Madera. Quetzali, Oscar Castillo ⁎, Mario García-Valdez, Alejandra Mancilla..(2017). A method based on Interactive Evolutionary Computation and fuzzy logic for increasing the effectiveness of advertising campaigns, Information Sciences 414 (2017) 175–186*

*Marion Garaus and Udo Wagner , Anna-Maria B¨ack.(2017). The Effect of Media Multitasking on Advertising Message Effectiveness, Psychology & Marketing, Vol. 34(2): )138–156 (February 2017*

*Martín-Santana J.D., E. Reinares-Lara , P. Reinares-Lara .(2017).Influence of radio spokesperson gender and vocal pitch on advertising effectiveness: The role of listener gender, Spanish Journal of Marketing – ESIC, Volume 21, Issue 1, February 2017, Pages 63-71*

*McKay-Nesbitt, J., V. Manchanda, R., C. Smith, M., & A. Huhmann, B. (2011). Effects of age, need for cognition, and affective intensity on advertising effectiveness. Journal of Business Research (64), 12–17*

*Moch, Nils (2013). Competition in fragmented markets: New evidence from the German banking industry in the light of the subprime crisis. Journal of Banking &; Finance, Volume 37, Issue 8, Pp 2908–2919*

*Moorthy, S. and Hawkins, S. (2005). Advertising repetition and quality perception. Journal of .Business Research, 58, 354– 360*

*Mucan. Burcu, Özeltürkay. Eda Yaşa.(2014).Social Media Creates Competitive Advantages: How Turkish Banks Use This Power? A Content Analysis of Turkish Banks through Their Webpages, Procedia - Social and Behavioral Sciences, Volume 148, 25 August 2014, Pages 137-145*

*Muda, M.; Musa, R.; Niana-Mohamed, R. and Borhan, H. (2014). Celebrity Entrepreneur .Endorsement and Advertising Effectiveness. Social and Behavioral Sciences, 130, 11-20*

*Oladele. Kehinde Joseph, Ogunnaike, Akinbode Mosunmola, Aka Deborah .(2016). Effective Advertising: Tool For Achieving Client-Customer Relationships, Researchjournali's Journal of Media Studies Vol. 2 | No. 1 February | 2016*

*Olsen, G., Pracejus, J. W., & O'Guinn, T. (2012). Print advertising: White space. Journal of .Business Research (65), 855–860*

*Park, M. (2006). The compensatory effects of pictorial and verbal information for haptic information on consumer responses in non-store shopping environments. Unpublished Ph.D, .Ohio, Ohio state University*

*Park, T. (2008). Effective design of advertisement in handheld devices (Doctoral dissertation .)PURDUE UNIVERSITY*

*Park, T. (2008). Effectiveness Design of Advertisement in Handheld Devices. Dissertation for .doctoral degree, Purdue University, West Lafyette India*

*Pergelova, A., D. Prior and J. Rialp, 2010 Assessing Advertising Efficiency Does the Internet Play a Role? Journal of Advertising, 39(3): 39-54*

*Pilelienė Lina Viktorija Grigaliūnaitė , (2017)," The effect of female celebrity spokesperson in FMCG advertising: neuromarketing approach ", Journal of Consumer Marketing, Vol. 34 - .Iss 3 pp*

*Rosemary J. Averya, Matthew D. Eisenbergb, Kosali I. Simon .(2012).The impact of direct-to-consumer television and magazine advertising on antidepressant use, Journal of Health Economics, Volume 31, Issue 5, September 2012, Pages 705–718*

*Selviaridis. Kostas, Martin Spring b, Luis Araujo b .(2013). Provider involvement in business service definition: A typology, Industrial Marketing Management 42 (2013) 1398–1410*

*Strebinger .Andreas, Xiaoling Guo, Ferdinand Klauser, Peter Grant-Hayd .(2018).Is Multi-Ethnic Advertising a globally viable strategy for a Western luxury car brand? A mixed-method cross-cultural study, Journal of Business Research 82 (2018) 409–416*

*Suvittawat, A., (2015). "Commercial Bank Service Quality: Chonburi Province, Thailand". .International Journal of Management Sciences and Business Research. 4(2): 2226-8235*

*Tafani. Eric, Elyette Rouxa, Rainer Greifeneder .(2018).In the mood for action: When negative program-induced mood improves the behavioral effectiveness of TV commercials, Journal of Business Research 84 (2018) 125–140*

*Te'eni-Harari, T. (2014). "Clarifying the relationship between involvement variables and .advertising effectiveness among young people". Journal of Consum Policy. 37: 183–203*

*Thomas, S. and Kohli, S.C. (2011). Can Brand Image Move Upwards after Sideways? A .Strategic Approach to Brand Placements. Business Horizons, 54 (1): 41-49*

*Yahyaoui . Hamdi, Own .Hala S..(2018). Unsupervised clustering of service performance behaviors, Information Sciences 422 (2018) 558–571*

*Yuksel, A., Yuksel, F. & Bilim, Y. (2010) "Destination Attachment: Effects on Customer Satisfaction and Cognitive, Affective and Cognitive Loyalty", Journal Of Tourism Management, Vol.31, pp: 274-284.*

# STUDY OF NATURAL ANTIOXIDANT INHIBITORS FOR CORROSION PROTECTION OF MILD STEEL IN ACIDIC MEDIUM

*Ms. Ibtihal Hassan AL-Bahrani, National University of Science and Technology, Sultanate of Oman.*

## ABSTRACT

Corrosion is a natural phenomena and it is a very big loss for structure as well as business; so corrosion minimization is one of the biggest areas of research. In this project; corrosion inhibition efficiency of berries juice extract was tested in mild steel sample against the sulfuric acid medium which was prepared from four types of berries; blackberries, blueberries, cranberries, and raspberries.

The experiment was designed by using Design Expert V 11 where immersed time, acid and berries juice extract concentrations were set as factors and a limited range was selected. The corrosion rate was set as a response. 20 runs were under taken and the experiments were done in the lab with maximum care.

The corrosion rate of 1020.374 mpy minimized to 597.659 mpy in presence of 5ml of extract and the efficiency of berries juice extract reached to 41.427%. The RSM gave 232.022 mpy corrosion rate as a minimum value and the optimum equation given by RSM was used for Genetic Algorithm analysis which gave 188.866 mpy corrosion rate at the optimizing conditions.

The surface and chemical characterization of the mild steel sample in presence and absence of berries juice extract were analyzed along with the relations between temperature, immersed time, acid and berries juice extract concentrations with response. The mixing tank was designed, calculations and drawing were presented.

**KEYWORDS**

Corrosion, Mild steel, Berries, Inhibitor, Weight loss, Sulfuric acid, RSM, SEM.

## INTRODUCTION

Mild steel is used in most of our daily life applications as well as in industrial applications. In other hands, mild steel has low corrosion resistance which corrodes easily once it reacts with the surrounding environment; corrosion is a phenomenon where the mild steel reacts with the deteriorates and surrounding environment. Corrosion of mild steel is a continuous and constant problem where it severe financial and life loses as it leads to loss in the product, efficiency of equipment and failures in some cases (Roberge,2012).

The corrosion phenomenon controlled by adding a small amount of component to the corrosive environment which is a corrosion inhibitor. There are organic and inorganic inhibitors. The inorganic inhibitors are used in limit amount as they are toxic such as nitrites, molybate, nitrate, silicate, chromate, and phosphate (Cafferty,2010). Now a day, the corrosion engineers looking for cheap, renewable, friendly, readily available and ecologically acceptable inhibitors which are organic inhibitors which comes from the plants such as antioxidant components consisting of tannins, saponins, flavonoids, alkaloids and amino acids. Organic inhibitors consist of some heteroatoms such as oxygen, nitrogen, multi-bound, and sulphur atoms with the conjugated system which allows the absorption in the metal surface. Those atoms attract the absorption of some components on the metal surface and that done by the interaction of metals and by the electron of the conjugated system or by forming a bond with the free electron of heteroatom which forms a layer in the metal surface, as a result, it decreases the corrosion rate (Roberge,2012).

Berries are small pulpy fruits which have up to 9.2 mmol of antioxidant per 100g of berries. Berries can be used as a corrosion inhibitor as they are rich in antioxidant. The studies show that the berries juice extract of blueberries, cranberries, raspberries, and blackberries rich with antioxidant where the antioxidant capacity of them reach to 10 mg/ 100 g of blueberries (Michalska&Lysiak,2015), 58 mg/100g of cranberries (Blumberg,J.,et al,2013), 26.2mg/ 100 g of raspberries (Bobinaite,R.,Viskelis&Venskutonis,2015), and 21 mg/100 g of blackberries (Kaume,L.,et al,2012).

### OBJECTIVES

To prepare anticorrosive extract (berries), to study corrosion in sample in presences and absences of the inhibitors, to determine the optimum conditions to inhibit corrosion and to design mixing tank.

## METHADOLOGY, DESIGN AND EXPERIMENTATION

### PREPARATION OF MILD STEEL SAMPLES

The mild steel sheets were cut into 2.5*3 $cm^2$ by using steel cutter and cleaned by surface cleaning machine to remove any corroded products in the metal surface then 4mm hole done by using drilling machine. Bunches for all metal samples (1-20) were made to differentiate between them. The drilled samples were washed by distilled water then rinsed with benzene and acetone respectively and dried with a hot air dryer. They were weighted using sensitive balance and kept in the desiccator until the time of usage.

### PREPARATION OF CORROSIVE MEDIUM

The corrosive medium of different concentrations of $H_2SO_4$ prepared by using 36.8 N of the sulfuric acid.

### PREPARATION OF BERRIES JUICE EXTRACT

137g, 209g, 282g, and 200g of raspberries, cranberries, blueberries, and blackberries respectively have been used to produce 309.5 ml of berries juice extract. The different types of berries washed by cold water, crashed in the electrical crasher and compressed mechanically by hand then filtered by cloth filter. The filtrate filtered again by using filter paper. The berries juice extract kept in the freezer at 0°C till the time of usage.

The efficiency of berries juice extract has been tested by running three experiments. The cleaned and weighted mild steel samples immersed in 0ml(blank), 2ml and 5ml of berries juice extract respectively with 50 ml of 1M $H_2SO_4$. After 2 hours; the samples have been taken out and cleaned by distilled water, benzene and acetone then dried by air dryer and weighted again. The corrosion rate and inhibitor concentration have been calculated by using equations (1) and (2) respectively.

$$Cr(mpy) = 22300 \times \frac{\Delta W}{\rho \times A \times t} \quad \textbf{equation(1)}$$

$$IE(\%) = \frac{\text{initalcorrosionrate-finalcorrosionrate}}{\text{initalcorrosionrate}} \times 100 \quad \textbf{equation(2)}$$

### DESIGN THE EXPERIMENT CONDITIONS

The experiments conditions have been designed by using Design Expert software-version 11 and following Response Surface Module(RSM); the range of the parameters have been given as inputs to the software which are inhibitor concentration (5-15 ml), immersed time (180-240 min) and acid concentration (1-2 M) and the response set as corrosion rate in mpy. Table(1) shows the experiment conditions given by the software.

**Table(1):** Designed runs given by RSM.

| Run.No | Factor(A): inhibitor, ml | Factor(B): Time, min | Factor(C): Acid concentration, M |
|---|---|---|---|
| 1 | 10 | 180 | 1 |
| 2 | 15 | 240 | 2 |
| 3 | 5 | 120 | 1 |
| 4 | 5 | 240 | 1 |
| 5 | 10 | 180 | 1.5 |
| 6 | 15 | 180 | 2 |
| 7 | 10 | 120 | 1.5 |
| 8 | 5 | 120 | 1.5 |
| 9 | 10 | 180 | 2 |
| 10 | 10 | 180 | 1.5 |
| 11 | 10 | 180 | 1.5 |
| 12 | 15 | 240 | 1 |
| 13 | 15 | 120 | 1 |
| 14 | 5 | 240 | 2 |
| 15 | 10 | 180 | 1.5 |
| 16 | 15 | 180 | 1.5 |
| 17 | 5 | 120 | 2 |
| 18 | 10 | 180 | 1.5 |
| 19 | 10 | 180 | 1.5 |
| 20 | 10 | 240 | 1.5 |

### APPLYING THE DESIGNED RUNS IN THE LABORATORY

The weighted mild steel sample has been hanged by using a plastic thread in a 50 ml of sulfuric acid in presence of berries juice extract of known concentration and contact time as per the software; the batch has been kept in the shaker. After the contact time the mild steel sample cleaned by using 120 grade emery paper to remove any corroded products, washed by distilled water, rinsed by benzene and acetone respectively and dried by hot air dryer, weighted by using sensitive balance. The corrosion rate in mpy was calculated by using equation(1) where the weight losses results have been used.

## STUDYING THE IMPACT OF TEMPERATURE IN CORROSION RATE

To study the impact of temperature in the corrosion rate; mild steel samples weighted and hanged for 2 hours in 50 ml of 1M sulfuric acid with 10ml of berries extract at different temperatures(30, 40, 50 and 60 °C) as shown in Table(2); each batch has been kept in the water bath shaker with heater to provide perfect contact between the solution and mild steel sample and to maintain the required temperature. Corrosion rate has been calculated by equation(1).

**Table(2):** Experiment conditions of temperature into corrosion rate.

| SI.NO | Temperature (°C) | Inhibitor concentration (ml) | Acid concentration (M) | immersed time, (hours) |
|---|---|---|---|---|
| 1 | 30 | 10 | 1 | 2 |
| 2 | 40 | 10 | 1 | 2 |
| 3 | 50 | 10 | 1 | 2 |
| 4 | 60 | 10 | 1 | 2 |

## SURFACE AND CHEMICAL CHARACTERIZATION ANALYSIS

The mild steel surface of blank, absence and presence of berries juice extract samples have been studied by using Scanning Electron Microscopy (SEM) and Energy Dispersive Spectroscopy (EDS).

## OPTIMIZATION STUDY BY GENETIC ALGORITHM

Matlab has been used for optimization where the genetic algorithm method has been followed. The actual equation form Design expert(RSM) has been used where "Optimtool" followed. The same limit of the conditions from RSM has been selected. The same previous procedure has been followed in the lab with maximum care by using those optimum conditions which given by the software, 1.001M of $H_2SO_4$ with 15 ml of berries juice extract immersed for 237.208 minutes.

## DESIGN OF MIXING TANK

Design of mixing tank is a part of the project where the industrial scale mixing tank has been designed, the tank will be formed from high resistance material such as S.S to be able to resist the corrosive medium of sulfuric acid which will be mixed with berries juice extract. It is a container which used to mix various material (heterogeneous materials) together to form homogenous materials even the different viscosity materials can be mixed easily which provides continuous process, good control, easy cleaning, and simplest construction. It can be made from S.S material which provides a smooth surface and it can be easily cleaned (Petropedia,2018).

Parameters to be considered to design mixing tank are length, torque, weight, stiffness and hydraulic loading.(Dynamixianc,2015). The important factors in design mixing tank are tank type, tank volume, specific gravity, process and environment.(Ibryamoval,Y.et al,2009)

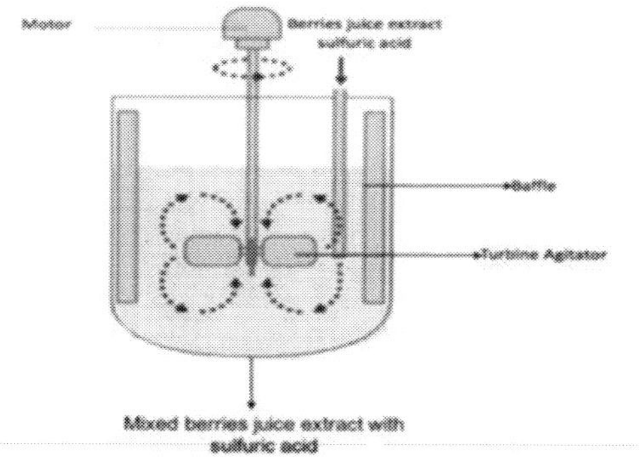


**Figure(1):** Mixing Tank(*Mixing Tank*,2018)

From the graph in Figure(2) Turbine agitator has been selected as it is suitable with the viscosity of the fluid and the tank volume. (Sinnot&Towler,2009)

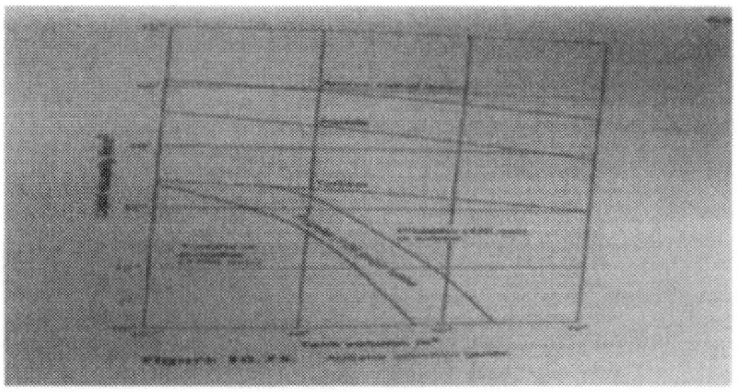

**Figure(2):** Graph of liquid viscosity Vs tank volume. (Sinnot&Towler,2009)

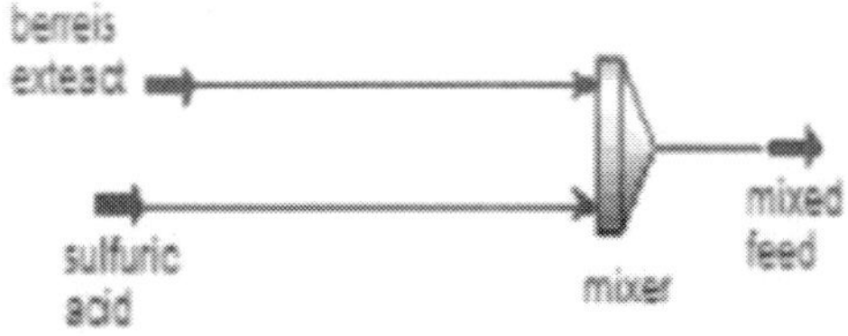


**Figure(3):** Result of Mixing Tank by Aspen Hysys.

**Table(3):** Calculated parameters for designing mixing tank

| | | |
|---|---|---|
| **Tank diameter** | m | 3.5 |
| **Tank Height** | m | 1.81 |
| **Across sectional area** | $m^2$ | 9.62 |
| **Volume of the tank** | $m^3$ | 17.5 |
| **Impeller diameter** | m | 1.16 |
| **Length of paddle** | m | 0.29 |
| **Width of paddle** | m | 0.23 |
| **power** | watt | 204085 |
| **Reynolds Number** | - | 136486.6391 |

## RESULTS AND ANALYSIS

### EFFECT OF BERRIES JUICE EXTRACT CONCENTRATION INTO CORROSION RATE AND INHIBITOR EFFICIENCY

As shown in Table(4) the corrosion rate in the absence of berries juice extract inhibitor reach to the maximum (1020.374mpy); the corrosion rate has been minimized to 898.117mpy and 597.659mpy in presence of 2 and 5 ml of berries juice extract inhibitor respectively which means that berries juice extract inhibits the corrosion as the concentration of the berries juice extract increases. As well as the inhibitor efficiency reaches to 41.427% which obtained at 5ml of berries juice extract inhibitor in 1M of sulfuric acid immersed for 2 hours. The adsorption of the organic compounds such as ascorbic acid that present in the berries juice extract is blocking the active sites on the mild steel samples as it is forming a protective film on the metal surface as what kumar.A,2012 mention in his studies so the corrosion rate keeps decreasing as the amount of the berries juice extract is increasing. (Kumar.A,2012) Those compounds have heteroatoms such as oxygen which is a center of adsorption. In addition; It forms a coordinate bond as it has the ability to transfer the electrons to the vacant d-orbital of mild steel from the heteroatoms. (Ostovari, L. et al,2009)

**Table(4):** Experimental results of testing the efficiency of berries juice extract.

| SI.NO | Inhibitor concentration (ml) | Corrosion rate (mpy) | Inhibitor efficiency (%) |
|---|---|---|---|
| 1 | 0 | 1020.374 | - |
| 2 | 2 | 898.117 | 11.976 |
| 3 | 5 | 597.659 | 41.427 |

### EFFECT OF IMMERSED TIME INTO CORROSION RATE AND INHIBITOR EFFICIENCY

The maximum corrosion rate reaches to 765.005mpy which obtained at 5ml of berries juice extract in the presence of 50 ml of 2 M of $H_2SO_4$ at 2 hours. Since the minimum corrosion rate of 232.022mpy obtained in presence of 5ml of berries juice extract with 50 ml of 1 M of $H_2SO_4$ immersed for 4 hours. The effect of immersed time into corrosion rate can be observed at low acid concentration and in the presence of berries juice extract, the corrosion rate is minimized at high immersed time. The decrease in the corrosion rate with the increase in immersed time is related to the adsorption of the corrosion product ($FeSO_4$) as shown in equation(3); According to Xianghong,L. et al ; the corrosion product of $FeSO_4$ becomes thicker with the immersed time, so it form a protective film (barrier) in the surface of mild steel (Xianghong, L. et al,2009).

$Fe_{(s)}+H_2SO_{4(aq)} \rightarrow FeSO_{4(aq)}+H_{2(g)}$ , **equation(3)**

## EFFECT OF ACID CONCENTRATION INTO CORROSION RATE

The effect of acid concentration into corrosion rate has been studied by using 1, 1.5 and 2 M of $H_2SO_4$ at different inhibitor concentrations and immersed time where the corrosion rate increase with the acid concentration at low inhibitor concentration and immersed time which reach to 765.005mpy at 2M $H_2SO_4$, 5ml of berries juice extract which immersed for 120 minutes as a maximum corrosion rate as shown in Table (3). In presence of the corrosive medium of $H_2SO_4$ with less amount of berries juice extract; the mild steel surface was exposed to $H_2SO_4$ medium where the corrosion reaction started as shown in equation(3) and $FeSO_4$ was formed with $H_2$ bubbles.

## EFFECT OF TEMPERATURE INTO CORROSION RATE

From Table(5) can be noted that the rate of corrosion increases with temperature where it reaches to maximum 2407.994mpy at maximum temperature of 60°C. According to Ambrish, S. et al, at high temperature the berries juice extract molecules were partially absorbed in the mild steel surface so it was exposed to the corrosive medium of $H_2SO_4$, as a result, increasing the corrosion rate (Ambrish,S. et al,2012).

**Table(5):** Results of studying the effect of temperature into corrosion rate

| SI.NO | Temperature,° C | Corrosion rate, mpy |
|---|---|---|
| 1 | 30 | 701.507 |
| 2 | 40 | 1666.404 |
| 3 | 50 | 2008.861 |
| 4 | 60 | 2407.994 |

## ANALYSIS BY RSM

Once the experimental results and corrosion rate calculated the RSM analysis has been done where the corrosion rate has been given to the software as a response. The response must be modified where the results fit with linear. The fit summary shows that the model F-value and P-value of 12.66 and 0.0001 respectively imply that the model is significant where A, B, C, and $B^2$ are the significant terms in this model only. 1.48 is a lack of fit P-value which indicates that the fit is not significant as there are pure errors.

0.5562 of the predicted $R^2$ is a reasonable agreement with 0.7105 of the adjusted $R^2$. The signal to noise ratio of 13.7252 of Adeq precision that the ratio grater than 4 is desirable. (Mycourses,2016)

The process variables are studied using ANOVA. The estimate coefficients in term of coded factors are A-Inhibitor, B-time, C-Acid and $B^2$. It is creating an accompanying condition for the sample with 1.0000 correlation coefficient. The coded and actual equations given by software are shown in equation(4) and (5) respectively.

- Corrosion rate(mpy)= 370.52-40.52(A)-126.97(B)+ 80.85(C) +66.58$(B)^2$ **equation(4)**
- Corrosion rate(mpy)= 1189.15487-8.10479(Inhibitor) -8.77421(Time) +161.69302 (Acid Conc)+0.018495 $(Time)^2$ **equation(5)**

The various relations can be studied from one factor graph; it can be noted that the corrosion rate increases with acid concentration and decreases with the increase in the amount of berries juice extract and immersed time as can be seen from the plots in Figure(3):Plot pf corrsoion rate Vs inhibitor concntration, Figure(4): plot of corrosion rate Vs acid concentration and Figure(5):Plot of corrosion rate Vs immersed time respectively.

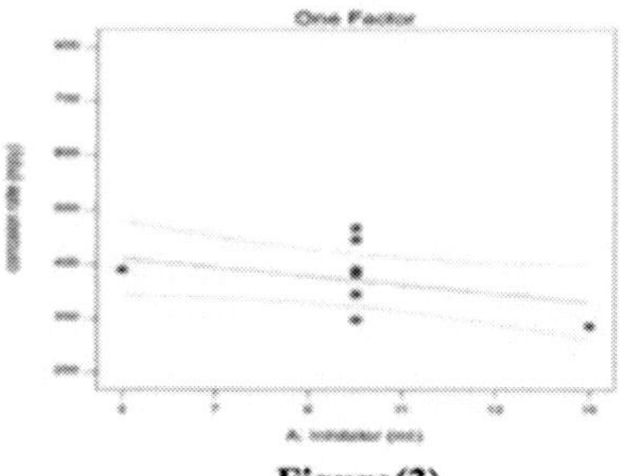


**Figure(3)**

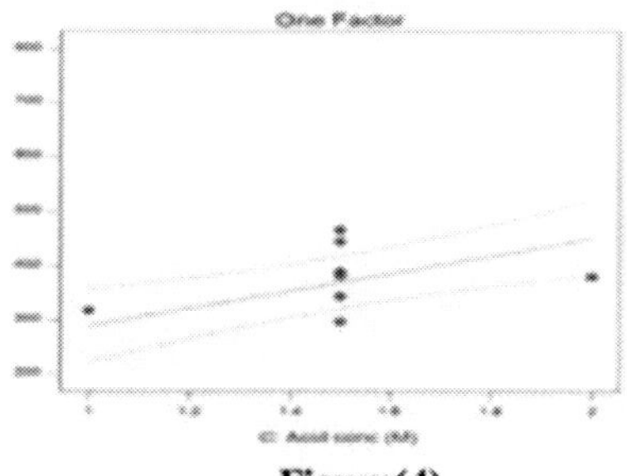


**Figure(4)**

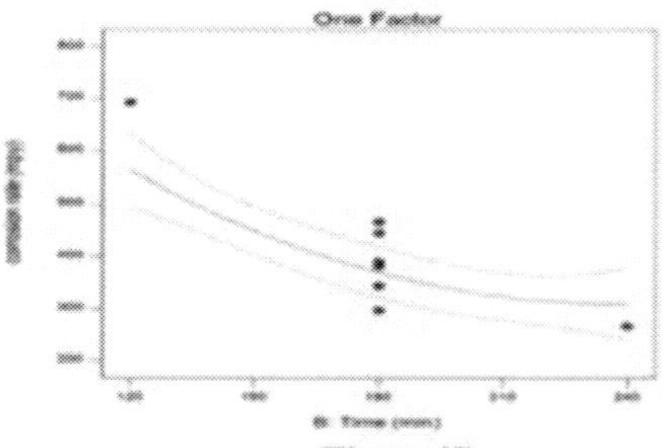


**Figure(5)**

The effects of two factors into the corrosion rate can be studied from the 3D plots as shown in Figure(6): 3D plot of corrosion rate Vs inhibitor concntration and acid concntration and Figure(7): 3D plot of corrosion rate Vs inhbitor concntration and immersed time.

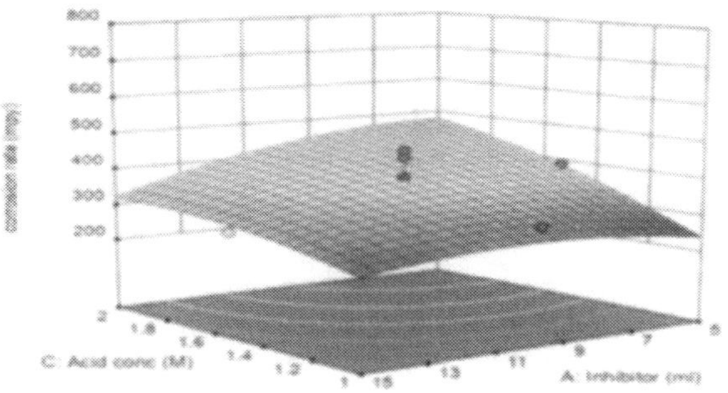


Figure(6)

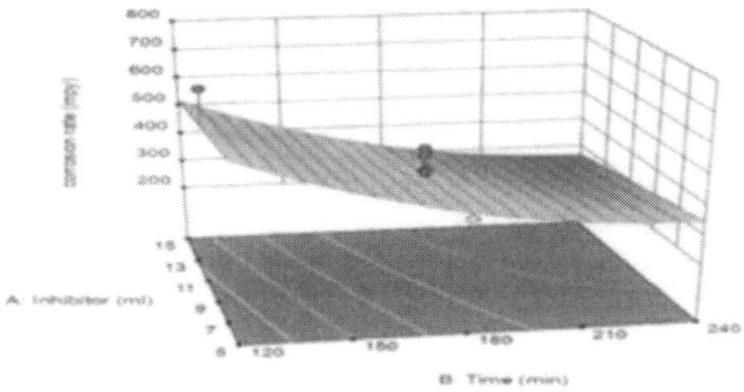


Figure(7)

The optimum conditions given by the RSM of 1M of $H_2SO_4$ with 5ml of berries juice extract inhibitor runs for 240 minutes gives 213.402mpy as a minimum corrosion rate but the experimental results at the same conditions gives 232.022mpy which means that they are too close as the suggested solution considered as an excellent optimum conditions.

**SEM ANALYSIS**

The SEM has been done for the blank mild steel sample for facilitating comparison as shown in Figure(8) which shows smooth mild steel surface compare to others. Rough surface in absence of berries juice extract is shown in Figure(9); where corrosion starting with the formation of $H_2$ bubbles as per the equation(3) which means that the surface of mild steel is exposed to the $H_2SO_4$ medium. In presence of 15 ml of berries juice extract, the surface of mild steel is protected by the adsorption of the organic compounds that are present in the berries juice extract where fewer cracks and pits can be noted and smooth surface is shown in Figure (10), which means that the mild steel surface is protected. (Raja, P. et al, 2008)

Figure(8)

Figure(9)

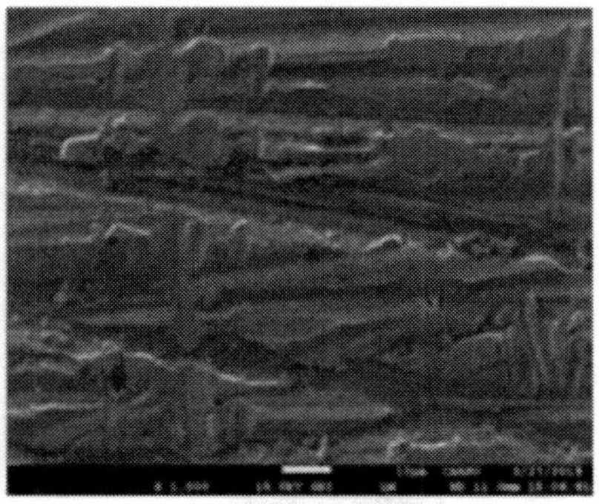

Figure(10)

**EDS ANALYSIS**

EDS has been used to study the chemical characterization or elemental analysis; Figure (11) shows the blank mild steel sample before being immersed in 1M $H_2SO_4$ in absence of berries juice extract where high Fe peak is shown, which indicates the presence of high amount of iron which reaches to 88 Wt% in the mild steel surface. Figure (12) shows the mild steel surface that has been immersed for 2 hours in 50 ml of 1M $H_2SO_4$ in absence of berries juice extract; the mild steel surface contains low amount of Fe (87.3 Wt%) compared to the blank mild steel sample. In Figure (13) the mild steel sample immersed for 2 hours in 50 ml of 1M $H_2SO_4$ in the presence of 5ml of berries juice extract, it shows high spark of Fe where the mild steel surface contains lower amount of Fe (84.8 Wt%) compared to Figure(11) and (12) which indicate lower iron dissolution rate in 1M $H_2SO_4$ medium containing berries juice extract (Alaneme, K. et al,2016). A high amount of oxygen and carbon are observed in presence of berries juice extract compared to the absence of berries juice extract. Si is shown in the absence and presence of berries juice extract which means that's it comes out to the surface during the reaction with sulfuric acid.

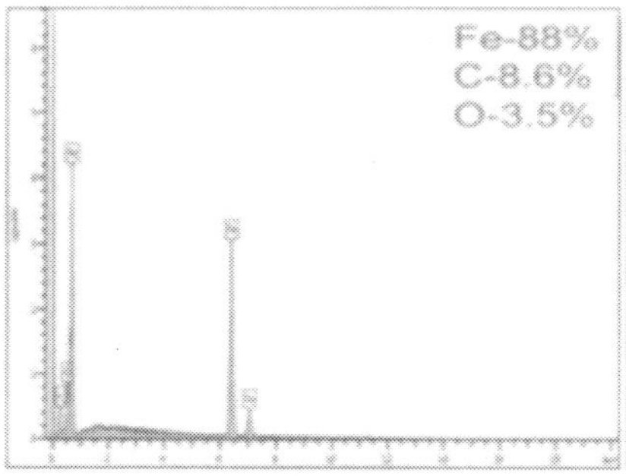


Figure(11)

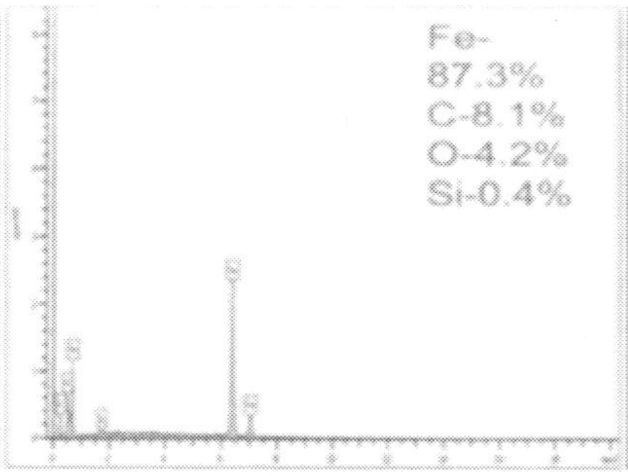


Figure(12)

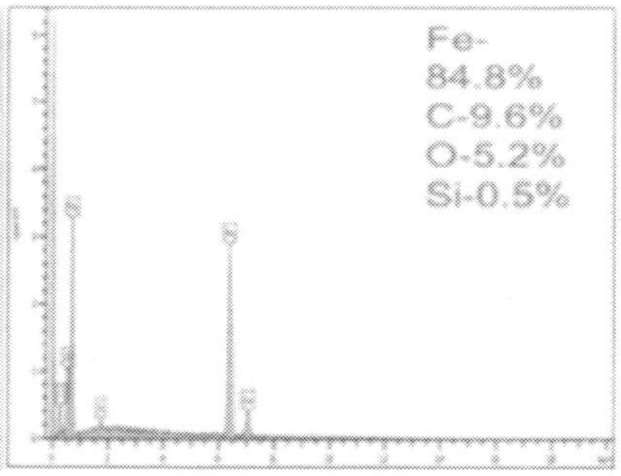


Figure(13)

**GENETIC ALGORITHM OPTIMIZATION**

The experimental result gives 288.367mpy as a corrosion rate. The software gives the minimum corrosion rate value of 188.866mpy from 50 runs as shown in Figure(14). A small difference can be noted as the genetic algorithm is not the exact fit with linear equations so the experimental and software results are too close to each other.

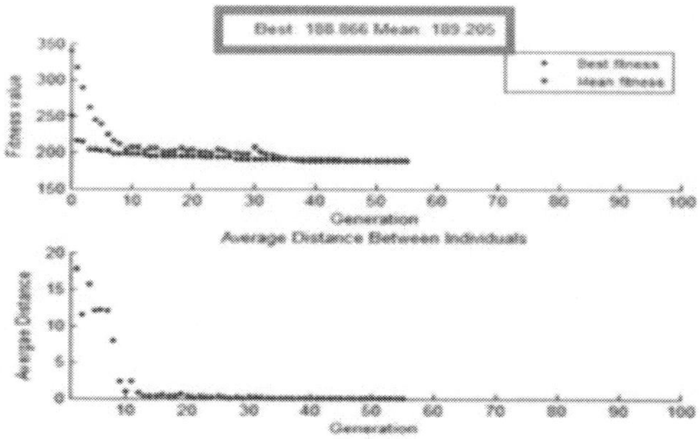


**Figure(14):** Genetic Algorithm optimization.

## CONCLUSION

In this project, all the objectives have been achieved by designing and experimenting.

Firstly; the anticorrosive extract (berries) has been prepared, the prepared extract which contains 4 types of berries namely blueberries, blackberries, cranberries, and raspberries inhibit the mild steel corrosion in sulfuric acid medium with 41.427% efficiency in presence of 5 ml of berries extract.

Secondly, the mild steel corrosion in the presence and absence of berries juice extract were studied where the corrosion rate reaches to 1020.374mpy in absence of berries juice. By using 5 ml of berries juice extract the corrosion rate minimized to 232.022mpy.

Thirdly, the optimum conditions to inhibit the mild steel corrosion were studied by Design Expert V11. At the optimum conditions of 1M of $H_2SO_4$. 5ml of berries juice extract, 240 minutes and at room temperature the corrosion rate reaches to 232.022mpy as a minimum value.

The actual equation obtained from Design Expert V11 was later optimized using Genetic Algorithm. Genetic Algorithm shows that at 1.001M of $H_2SO_4$, 15 ml of berries juice extract and 237.208 minutes at room temperature, the corrosion rate is minimized to 188.866mpy.

Finally, mixing tank has been designed with 1.18m height and 2.5m diameter and a turbine agitator has been selected.


## ACKNOWLEDGEMENT

I would like to thank god for giving me patience and power to work in my most important part of my life, my father, mother, my brother and my friends for their help and support. I would also like to express my sincere gratitude to Dr. Priy Brat Dwivedi for his support and help in developing my project idea, Mr. M.J.Varghese for his help in RSM analysis, Mr. Hatem Yazidi and Dr. Moammar Ali AlTubi for providing me the technical knowledge in Matlab optimization, Mr.Mohammed Salim Adeeb in providing me the technical help in the laboratory, Ganesh Nagraj Patil, Dr.Lakkim Setty

Rao, and Dr.Santosh Madhav Walke for their help in the calculations of design mixing tank. Last but not the least, I would like to thank National University of Science and Technology-College of Engineering for giving me this opportunity to complete the project in their laboratories and supporting me by providing me with the required chemicals.

# REFERENCES

**1.**Alaneme,K. et al .,2016.Corrosion Inhibitory Properties of Elephant Grass(Pennisetum Purpureum) Extract Effect on Mild Steel Corrosion in 1M HCl Solution :*Alexandria Engineering Journal*.[e-journal]. 55 (8). p. 1069-1076. Available from: https://www.sciencedirect.com/science/article/pii/S1110016816300242.[Accessed: 8th May 2019].

**2.**Ambrish,S. et al .,2012. The Effect of Environmentally Bening Fruit Extract of Shahjan(Moringa Olifera) on the Corrosion of Mild Steel in Hydrochloric Acid Solution:*Journal of Chemical Engineering Communications*.[e-journal]. 199 (6). p. 63-77. Available from:https://www.tandfonline.com/doi/abs/10.1080/00986445.2011.570390.[Accessed: 6th May 2019].

**3.**Blumberg,J.,et al.,2013.Carnberries and Their Bioactive Constituents in Human Health:*Journal of Advances in Nutrition*.[e-journal]. 12 (7). p. 618-632. Available from: https://academic.oup.com/advances/article/4/6/618/4595566.[Accessed: 1th May 2019].

**4.**Bobinaite,R.,Viskelis,P.&Venskutonis,P.,2015.Chemical Composition of Raspberry Cultivars:*International Journal of Molecular Sciences*.[e-journal]. 16 (8). p. 18642-18663. Available from: https://www.researchgate.net/publication/306158828_Chemical_Composition_of_Raspberry_Rubus_spp_Cultivars.[Accessed: 1th May 2019].

**5.**Cafferty,E.,2010.*Introduction to Corrosion Science*.United States:Springer.
Roberge,P.,2012.*Handbook of Corrosion Engineering*.2nd edition.United States:McGraw-Hill company.
**6.**Dynamixianc,2015.*Optimal Mixer Design* .[Online]. Available from: https://www.dynamixinc.com/optimal-mixer-design. [Accessed: 14th December 2018]

**7.**Ibryamoval,Y. et al .,2009.Design of Mixers for Chemical and Biochemical Industry: *Trakia Journal of Sciences*.[e-journal]. 8 (6). p. 6-11. Available from: http://tru.uni-sz.bg/tsj/Vol8.Suppl.3.2010/Y.Ibryamova.pdf.[Accessed: 13th December 2018].

**8.**Kaume, L. et al.,2012. The Blackberry Fruit A Review on Its Composition and Chemistry, Metabolism and Bioavailability and Health Benefits:*Journal of Agricultural and Food Chemistry*.[e-journal]. 37 (7). p. 5716-5727. Available from: https://www.ncbi.nlm.nih.gov/pubmed/22082199.[Accessed: 6th May 2019].

**9.**Kumar.A,2012. Inhibition of Mild Steel Corrosion in HCl Solution Using Pipali(Piper Longum)Fruit Extract:*Arabian Journal for Science and Engineering*.[e-journal]. 38 (9). p. 85-97. Available from: https://link.springer.com/article/10.1007/s13369-012-0409-9.[Accessed: 9th May 2019].

**10.**Michalska,A.& Lysiak.G .,2015.Bioactive Compounds of Bluberries:*International Journal of Molecular Sciences*.[e-journal]. 16 (8). p. 18642-18663. Available from: https://www.ncbi.nlm.nih.gov/pmc/articles/PMC4581264/.[Accessed: 1th May 2019].

**11.***Mixing Tank*,2018.[Online]. Available from: https://www.conceptdraw.com/examples/agitator-tank-symbol.[Accessed: 19th November 2018].

**12.**Mycourses,2016. *Response Surface Modelling-RSM*.[Online]. Available from:https://mycourses.aalto.fi/pluginfile.php/364340/mod_resource/content/1/BOS_Lecture3_2016.[Accessed: 4th May 2019].

**13.**Ostovari, L. et al.,2009.Corrosion Inhibition of Mild Steel in 1M HCl Soluition by henna extract:*Journal of Corrosion Science*.[e-journal]. 32 (9). p. 1935-1949. Available from: https://www.sciencedirect.com/science/article/pii/S0010938X09002157. [Accessed: 6th May 2019].

**14.**Petropedia,2018. *Mixing Tank*.[Online]. Available from: https://www.petropedia.com/definition/7742/mixing-tank.[Accessed: 11th December 2018].

**15**.Raja,P. et al.,2008.Inhibitive Effect of Black Pepper Extract on the Sulphuric Acid Corrosion of Mild Steel:*Journal of Materials Letters*.[e-journal]. 62 (9). p. 1977-2979. Available from: https://www.sciencedirect.com/science/article/abs/pii/S0167577X08001316.[Accessed: 8th May 2019].

**16.**Sinnot,R.&Towler,G.,2009.*Chemical Engineering Design*.5th edition.India:Elsevier.

**17.**Xianghong,L. et al.,2009.Inhibition Effect of 6-benzylaminopurine on the Corrosion of Cold Rolled Ssteel in $H_2SO_4$ Soluition:*Journal of Corrosion Science*.[e-journal]. 51 (9). p. 620-634. Available from: https://www.sciencedirect.com/science/article/pii/S0010938X09000080. [Accessed: 6th May 2019].

# 590434 Verification of Firefighters' Heuristics and Predictive Model Development through Big Data Analysis

**Abstract:**

The heuristics accumulated in the field activities of firefighters are reviewed through big data analysis of fire occurrences and emergency responses in Gyeonggi-do Province, Korea and expanded as predictive models to be utilized for proper fire prevention activities according to time, day, and target through quantitative modeling. Empirical rules with high sympathy are collected through direct interviews with firefighters. Among these, three types of high-priority rules are compared and analyzed for big data such as fire occurrence and damage history in Gyeonggi-do in 2018. Big data comparison analyses are conducted, including the number of fires and damages that occurred in Gyeonggi-do in 2018. Furthermore, fire occurrence patterns by region, day of the week, time of day, and building type are derived. In addition to using empirical rules that have been validated through research, relatively inexperienced firefighters also can make better decisions by relying on refined quantitative predictive models replacing empirical rules, including local government and time-based factors that reflect big fire occurrence data.

# CONVERTING SHELTER-IN-PLACE LOCATIONS TO ZERO VULNERABILITY SAFE HAVENS

J. A. Rau
*MineARC Systems America, LLC*
*4850 W. Ledbetter Drive*
*Dallas, Texas 75236*
*USA*

A. Young
*MineARC Systems, Pty Ltd*
*274 Welshpool Rd*
*Welshpool, WA 6106*
*Australia*

M. McDermott
*High Performance Building Solutions*
*7415 Folger Dr.*
*Charlotte, NC 28226*
*USA*

# CONVERTING SHELTER-IN-PLACE LOCATIONS TO ZERO VULNERABILITY SAFE HAVENS


## ABSTRACT

While it is not uncommon to find shelter-in-place or safe haven locations designated across chemical facilities, what varies is the level of protection offered at these sites. Functionality can range from merely a muster point (shelter-in-place) to a positively pressurized room (safe haven). Either way, when determining the room's acceptable performance criteria, the industry generally applies ERPG-3[1] or a vulnerability factor. An acceptable vulnerability factor is often deemed the probability of a fatality being less than one percent.

In 2017, MineARC Systems and High Performance Building Solutions (HPBS) began working with a leading energy company to convert five shelter-in-place spaces to safe havens, which included additional life support systems that would ensure zero percent vulnerability. The objective of this paper is to examine the process to harden the space against contaminated air intrusion and what critical life support systems are essential for converting a shelter-in-place location to a safe haven.


## INTRODUCTION

Emergency shelter-in-place (SIP) forms an integral part of any plant's broader emergency response plan. Explosions, fires, and the release of smoke or other toxic gases are some of the types of incidents that can occur despite high levels of planning and safety precautions[2].

Numerous government agencies and other organizations recommend shelter-in-place as a measure to reduce harm in the event of a chemical release. Any building may be used as a temporary measure to reduce health risk from exposure to the toxic materials by simply closing windows and doors and turning off ventilation fans and air conditioning/heating systems.

Three factors govern the effectiveness of a SIP location: 1) the tightness of the building or leakage rate, 2) the concentration of the toxic gas, and 3) the release duration. The tightness of the building is primarily based on building construction; however, temperature differential and wind speed also play a factor. Identification of the maximum toxic concentration occurs during a Facility Siting Study (FSS). FSSs are a hazard analysis that defines Maximum Credible Event (MCE) scenarios using a consequence-based approach to produce a quantitative risk assessment. Determining potential consequences is a necessary step in the process of developing a comprehensive safety plan. The FSS determines the expected release duration based on stored inventories of toxic materials. The higher the toxic concentration and longer the release duration, the less effective sheltering in place becomes.

While occupants may feel safe sheltering-in-place from the external toxic hazard, typical buildings have high leakage rates. A typical wood framed constructed building can have 3-5 air changes per hour (ACH50[3]), rendering it ineffective against most chemical releases. Additionally, sealing to an acceptable airtightness can be time consuming and expensive and even with professional air sealing it is very difficult to air seal existing structures to an acceptable tightness.

A case study for the conversion of five shelter-in-place spaces to zero percent vulnerability safe havens is presented. There are considerations for effectively sealing existing infrastructure and ensuring positive pressure to ensure contaminants do not infiltrate safe haven locations along with the critical element of life support. The critical elements of life support that are of primary importance to human survival in a safe haven are:

- Positive pressure
- Breathable air supply
- Supplemental $O_2$
- Removal of accumulated $CO_2$

---

[1] Emergency Response Planning Guidelines (ERPGs) are values developed by the AIHA Guideline Foundation's Emergency Response Planning Committee. ERPG-3 is the maximum airborne concentration below which it is believed nearly all individuals could be exposed for up to 1 hour without experiencing or developing life-threatening health effects.

2 The CSB is an independent federal agency charged with investigating industrial chemical accidents. Over the last 10 years they have completed investigations into 15 separate chemical release incidents.

[3] ACH50 is the number of times the air volume in a building changes per hour at 50 Pascal of pressure. Commercial air leakage rates are commonly reported at 75 Pascals, but this leakage study was conducted at 50 Pascals based on the targets included in the FSS report.

- Cooling and dehumidifying

## CASE STUDY

In 2017 MineARC Systems and its project partner HPBS began working alongside a major energy company to convert five shelter-in-place locations into zero percent vulnerability safe havens.

The process started with the energy company completing its FSS, which included plume modeling to define the chemical release concentrations at each building location along with ideal shelter-in-place locations[4] and internal minimum positive pressure[5]. This information was used by MineARC to calculate indoor concentrations and occupant vulnerability to highlight that shelter-in-place alone would result in the vulnerability of greater than zero percent.

From there, HPBS conducted air infiltration testing following ASTM E779-10 to determine an air leakage baseline (see Table 6) for rooms that were designated to be sealed (pre-weatherization). A site assessment was then completed to review each location and determine its suitability for sealing and conversion to a safe haven. Firstly, the measured baseline infiltration rates were compared to the FEMA 453 estimates based on construction type[6] (see Figure 2). All surfaces of the room were examined for ease of conventional air sealing and mechanical equipment sealing. Rooms that included false floors, large amounts of electrical conduit, cabinetry, computer servers or windows etc. were deemed difficult to seal (i.e. control rooms). Secondly, the building or rooms open floor space was checked to ensure it could accommodate the specified number of people in an emergency. FEMA 453 section 1.6 recommends 20ft$^2$ per person for durations of less than 24 hours. With the FSS recommending 4 hours, it was deemed acceptable to have an unrestricted floor space of 6ft$^2$ per person.

Manual weatherization (air sealing) of the rooms included installing a cleanroom suspended ceiling system with gaskets and tiles that met the air permeance rate for an air barrier material, cleanroom airtight light fixtures with gaskets, sealing around electrical and plumbing, air sealing all visible air pathways to make sure the interior gypsum board was functioning as the air barrier, installing low-leakage HVAC dampers, installing an emergency shut off for the mechanical system and adding door weather-stripping and sweeps. Post-weatherization air leakage tests were conducted to determine the air infiltration rate of the SIP spaces after completion of the manual air sealing, cleanroom suspended ceiling system and HVAC work. The weatherization air sealing methods improved the air barrier integrity for all the proposed SIP spaces by between 57% and 65% (refer to Table 6).

Aerosol-based envelope sealing was used to reduce leakage rates even further. AeroBarrier® is a building envelope sealing technology that simultaneously measures and seals buildings. AeroBarrier® works by forcing an atomized sealant into orifices under pressure gradually sealing leaks. Rooms were sealed during night shift with personnel able to re-access the rooms by morning. The AeroBarrier® air sealing methods improved the air leakage for all the SIP spaces between 7% and 88% over the post manual weatherization leakage rates (refer to Table 7). AeroBarrier® is not effective for sealing large holes so manual air sealing is required for certain locations prior to using AeroBarrier®.

MineARC engineers once again calculated infiltration concentration rates along with the vulnerability rate. As the probability of a fatality was still higher than zero, MineARC then specified its AirBANK Pressure System and necessary carbon dioxide and oxygen control equipment.

## SAFE HAVEN LIFE SUPPORT EQUIPMENT

### Primary Life Supporting Atmosphere

A safe haven needs to provide a reliable breathable atmosphere independent of external toxic gas concentrations. Human breathing consumes oxygen ($O_2$) and expels carbon dioxide ($CO_2$), carbon monoxide (CO), water vapor, and heat. At complete rest, an average person exhales air at 17% $O_2$ and 3.2% $CO_2$. This breathing rate results in a 'resting' $O_2$ uptake of 0.3L/min of $O_2$ and a resting $CO_2$ discharge of 0.24L/min (Brake, Fellows, & Bates, 1999). Taking into consideration the above values, a person in a closed system such as a shelter-in-place succumbs to $CO_2$ poisoning well before $O_2$

4 Shelter-in-place location is beyond the scope of this paper, but should be determined based on the chemical hazard, facility layout, and location of personnel.

5 Positive pressure requirements are based on average wind speed at the facility. As wind speed increases so does the SIP leakage rate.

6 The FEMA 453 estimates are intended to estimate filter sizing when ASTM E779 air leakage testing is not conducted.

asphyxiation. Venter et al. (1998) found an average oxygen consumption for 12 entrapped individuals of 0.44 liters per minute, at STP, over 24 hours (including sleeping). A figure of 0.5 liters per minute over a non-sleeping entrapment period of 12 hours is a prudent, conservative design value. This corresponds to a metabolic rate of about 160W or 80W/m$^2$ for a typical person with a body skin area of about 2m$^2$, which is equivalent to a breathing rate of about 12.5 liters per minute of fresh air.

Consequently, life support systems are needed within a safe haven that can ensure positive pressure keeps contaminants outside, along with replacing the $O_2$ consumed, and removing exhaled $CO_2$, moisture, and heat to maintain the atmosphere within acceptable threshold limits.

Positive Pressure

Positive pressure is integral to prevent the ingress of airborne contaminants that may compromise the breathable atmosphere for occupants. Establishing a Class 1 Safe Room (FEMA 453) requires the sufficient release of filtered air from an external source, to form a positively pressurized zone.

Buildings are typically constructed to provide natural or forced ventilation to allow permanent occupancy and potentially combined with large exposed windows and surfaces which can present challenges for ensuring protection during a chemical release. When determining adequate positive pressure within a space, several structural factors need to be considered, including:

- Shelter volume
- HVAC (Heating, Ventilating, and Air Conditioning)
- Door and window tightness
- Existing opening such as vents and elevator shafts
- Indoor and outdoor temperature
- Wind speed and building orientation
- The external screening of the structure

Building design and multiple ingress points suggest the conversion of a smaller interior room within a building can provide a higher level of protection, than the whole building, as it is less subject to wind and buoyancy forces which induce the infiltration of external hazards.

Breathable air from compressed air cylinders venting directly into the SIP forms a positive internal pressure. Determining the amount of air required to reach the necessary level of positive pressure involves measuring the areas internal volume and the relationship between the internal and external atmospheric conditions.

After reaching required ACH (air changes per hour) or tightness, a positive-pressure management system must be sized for the expected shelter duration.

Breathable Air Supply

If the shelter-in-place is airtight, a supply of breathable air is necessary to support occupants over the required shelter duration. Unventilated SIPs that are tightly sealed cannot be occupied for long periods without the risk of high carbon dioxide levels. A secondary source of breathable air must be circulated within the shelter to meet occupancy over the required amount of time.

Clean, breathable air is supplied to a chemical, biological and radiological (CBR) safe room in a few ways: by trapping air inside the room and minimizing the air exchange, passing contaminated air through a filter (not practical for high concentration chemical releases or for chemicals that cannot be filtered) to purify it as it is supplied to the room, and the supplemental supply of oxygen through high pressure compressed air cylinders.

The air supply must contain enough oxygen in its mix, and not contain contaminants like carbon monoxide, carbon dioxide, water, or any other toxins that poison safe haven occupants during entrapment.

Table 1 lists international standards for flowrates for breathable compressed air for self-contained breathing apparatus (SCBA), supplied air respirators (SAR) and refuge chambers.

Table 1 – Standards for acceptable air supply requirements

| Air supply requirements | Relevant Standard |
|---|---|
| Open-circuit positive-pressure SCBA, including all components and accessories except the air cylinder (shell) must resist the permeation and penetration of HD and GB chemical agents when tested for a period of 6 hours on an upper-torso manikin connected to a breathing machine operating at an airflow rate of 40 liters per minute (lpm), 36 respirations per minute, 1.1 liters tidal volume. | NIOSH 42 CFR Part 84 Respiratory Protective Devices |
| The flow rate should be at least 85L/min (3cfm) per person to remove expired CO2 and CO but at least 255L/min (9cfm) to provide adequate cooling and maintain comfortable conditions. The air inflow is assumed to be at a temperature of 26.5°C and 65% relative humidity | ITA Report No. 14 – Guideline for the provision of refuge chambers in tunnels under construction, May 2014 |
| For manually operated blower type air-hose respirators, the blower shall be capable of delivering air to the space enclosed by the facepiece at a flow rate of at least 85 L/min. | AS/NZS 17:16:2003 Respiratory Protective Devices |

The NIOSH standard of 40L/min is for SCBA apparatus (i.e. mouthpiece or face mask) where dilution of $CO_2$ is not a concern. Conversely, the ITA and Australian standards specify 85L/min and are concerned with both oxygen consumption and carbon dioxide dilution. It is important to note that three times the amount of air is required to maintain comfortable conditions inside a refuge chamber, when compared to simply breathing (Health & Safety Executive UK, 2006).

Unless the positive pressure system is supplying breathable air at a rate that maintains oxygen and compressed gas levels within acceptable ranges (dependent on safe haven occupancy and duration), a $CO_2$ scrubber and supplementary oxygen supply are required.

Supplemental $O_2$

$O_2$ is an essential requirement for cellular activity, and therefore, the absence of $O_2$ for relatively short periods results in death. Humans within an enclosed environment consume $O_2$ and if not replaced, would result in an irrespirable atmosphere. An $O_2$ concentration of less than 11% poses a high risk of death.

Retaining the oxygen levels between acceptable limits is critical. Various international standards outline acceptable $O_2$ limits for a breathable environment; examples are shown in Table 2.

Table 2 - Standards for allowable $O_2$ levels

| Minimum $O_2$ Content (% by volume) | Maximum $O_2$ Content (% by volume) | Relevant Standard |
|---|---|---|
| 18.5% | 23.0% | Mines Safety and Health Administration 30 CFR Parts 7 and 75 Refuge Alternatives for Underground Coal Mines; Final Rule |
| 19.5% | 23.5% | Occupational Safety and Health Administration 29 CFR 1910.146 Permit-Required Confined Spaces for General Industry |
| 19.5% | 23.5% | Australian Standard AS 2865–2009 Safe working in a confined space |

OSHA and Australian Standards for working in a confined space specify an acceptable lower limit of 19.5%, which is consistent with many other standards around the world. The slightly lower MSHA limit of 18.5% is based on a person at rest, not working, therefore, not likely to have trouble in breathing; conditions that are likely to be present in a refuge chamber.

$O_2$ itself is not flammable but supports and accelerates combustion. OSHA defines $O_2$ enriched mixtures or atmospheres as those containing more than 23.5% $O_2$ by volume (OSHA, 2013). In $O_2$-enriched atmospheres, the reactivity of $O_2$ significantly increases the risk of ignition and fire.

Within a safe haven, pure $O_2$ must be utilized in conjunction with a $CO_2$ removal system.

Removal of Accumulated $CO_2$

Protection within a safe haven is substantially increased by filtering $CO_2$ from the air. $CO_2$ is being produced continuously by the cells of our body and is a byproduct of respiration. The inability to expel this gas or an environment with high concentrations of $CO_2$ leads to $CO_2$ poisoning. Breathing air with more than 3% $CO_2$ can induce dizziness, headaches, and vision disturbance. Prolonged exposure at greater than 6% could result in unconsciousness and death. Listed in Table 3 are examples of international standards of satisfactory $CO_2$ removal requirements for a breathable environment.

Table 3 – Acceptable $CO_2$ removal requirements

| $CO_2$ Removal Requirements | Relevant Standard |
|---|---|
| $CO_2$ removal components shall remove $CO_2$ at a rate of 1.08 $ft^3$ (30.5 L)/h/person. | Mines Safety and Health Administration 30 CFR Parts 7 and 75 Refuge Alternatives for Underground Coal Mines; Final Rule |
| OSHA has established a Permissible Exposure Limit (PEL) for CO2 of 5,000 parts per million (ppm) (0.5% CO2 in air) averaged over an 8-hour workday (time-weighted average or TWA.) | Occupational Safety and Health Administration |
| 5.2.8 Atmospheric monitoring. Atmospheric monitoring There should be the capability to monitor temperature continuously, and for carbon monoxide, carbon dioxide, and oxygen in the chamber during use in both externally supported or standalone modes. The concentration of carbon dioxide should not exceed 10000 ppm. | ITA Report No. 14 – Guideline for the provision of refuge chambers in tunnels under construction, May 2014 |

The $CO_2$ that is exhaled by the occupants within the safe haven must be removed by, constant flushing by an external breathable air source (e.g. compressed air cylinders used for positive pressure) or by the regeneration of the internal atmosphere with a $CO_2$ absorbent (scrubber).

Different regulations around the world, state varied human production rates between 30–50L per hour per person. When considering an acceptable production rate; it must be noted that the rate of breathing is primarily triggered by the $CO_2$ content of inspired air with higher $CO_2$ levels triggering faster respiration rates (Brake, Fellows, & Bates, 1999) (i.e. if the maximum allowable $CO_2$ concentration inside the safe haven is 1%, a lower production rate is acceptable compared to a 3% $CO_2$ limit).

If $CO_2$ removal mechanisms are inadequate, the occupants can develop $CO_2$ toxicity by re-breathing their own exhaled $CO_2$.

Cooling and Dehumidifying

Thermal tolerability with a safe haven should be addressed during the design phase. Assumptions must be made in preparation for safe haven design in regard to the number of personnel, activity level, available hydration and heat production. The minimum airflow to maintain a comfortable environment is set by taking these assumptions into account as well as the input temperature and humidity of supplied air.

The healthy human body maintains its internal temperature around 37°C. An essential requirement for continued normal body function is that the deep body temperature is maintained within the narrow 1°C limit (Epstein, Moran, 2006). A change of body temperature exceeding 1°C occurs only during illness or when environmental conditions surpass the body's ability to cope with extreme temperatures.

When a person is subject to extreme heat and humidity over some time, the person may experience a very rapid progression of symptoms or exhibit a few of the less significant symptoms before falling victim to more severe forms of heat-related illness.

The acceptable cooling requirements for underground refuge chambers are defined in various international standards, examples of which are listed in the below table.

Table 4 – Standards for acceptable cooling requirements

| Cooling Requirements | Relevant Standard |
|---|---|
| Simulated emergencies, in which a full complement of people has occupied a refuge chamber for a significant period, indicate that humidity and temperature can increase very rapidly to potentially heatstroke-inducing levels. Refrigerated air conditioning is strongly recommended for both externally supported and stand-alone refuge chambers to counter this potentially dangerous problem. Inevitably, this places a heavy demand on the stand-alone power supply, but there are systems available that can cope, and at an acceptable cost. | Department of Mines, Industry Regulations and Safety, 2005 Refuge Chambers in Underground metalliferous mines – guideline: Safety and Health Division, Department of Industry and Resources Western Australia, 32 pp |
| Apparent temperature in the fully occupied refuge chamber must not exceed 95 °F. | Mines Safety and Health Administration 30 CFR Parts 7 and 75 Refuge Alternatives for Underground Coal Mines; Final Rule |
| It should be possible to maintain an internal temperature not exceeding 30 °C wet bulb globe temperature for at least 24 hours | ITA Report No. 14 – Guideline for the provision of refuge chambers in tunnels under construction, May 2014 |

MSHA regulations specify a maximum internal apparent temperature[7] of 95°F (35°C). NIOSH specifies an equivalent maximum wet bulb temperature of 28.9°C (NIOSH, 2007).

The temperature maintained inside the safe haven is determined by the metabolic heat liberated by the occupants, dissipated to the external environment by the heat transfer characteristics of the room or building. The external temperature primarily influences the heat transfer while the humidity is the combination of moisture produced by the occupants (via exhaled air and sweat) and water liberated by $CO_2$ scrubbing chemicals.

It is important to remember that in most cases, before sheltering-in-place, the air conditioning system will need to be shut down (unless it is a split system). For short durations, artificial cooling is not a critical requirement unless external temperatures are extraordinarily high, and/or the safe haven's volume is small compared to its occupancy.

## INFILTRATION TESTING

### Background

High Performance Building Solutions Inc. (HPBS) was contracted to perform an air infiltration study for spaces identified as candidates to be used as shelter-in-place (SIP) in the event of an accidental chemical release. This SIP strategy uses the existing structure and its indoor atmosphere to separate individuals from a hazardous outdoor atmosphere temporarily.

A series of pre-weatherization air leakage tests were conducted to determine the baseline air infiltration rate of potential SIP spaces. Manual air sealing, a cleanroom suspended ceiling system and low-leakage HVAC dampers were installed to reduce the overall air infiltration of the SIP spaces and to close off air leakage pathways from non-protected spaces in the building through the HVAC system. Post-weatherization air leakage tests were conducted to determine the air infiltration rate of the SIP space after the completion of weatherization and HVAC work. The SIP spaces were additionally air sealed using the proprietary AeroBarrier® system to further reduce infiltration rates.

### Scope

Air sealing work was completed, and low-leakage HVAC dampers were installed to reduce the overall air infiltration of the SIP spaces and to isolate the HVAC from non-protected spaces in the building. Post-weatherization air leakage tests were conducted to determine the air infiltration rate of the SIP space after the completion of air sealing, suspended ceiling and HVAC work. AeroBarrier® air sealing was conducted on SIP spaces as a pilot project on September 18 through September 21, 2017. Post AeroBarrier® air leakage tests were conducted on September 21 and September 22, 2017, to determine the air infiltration rate of the SIP space after the completion of the AeroBarrier® sealing.

7 Apparent temperature is the temperature equivalent perceived by humans, caused by the combined effects of air temperature and relative humidity.

Following ASTM E 779-10, pressurization air leakage tests were conducted on the following spaces recommended in the Facility Siting Study (FSS).

Table 5: SIP specifications

| SIP Space | Occupancy | Floor Area (Square Feet) | Volume (Cubic Feet) |
|---|---|---|---|
| SIP Space 1 | 40 | 990 | 10,481 |
| SIP Space 2 | 6 | 955 | 8,210 |
| SIP Space 3 | 5 | 391 | 3,917 |
| SIP Space 4 | 10 | 1,697 | 16,968 |
| SIP Space 5 | 40 | 293 | 2,637 |

Air Leakage Test Results

Positive pressure air leakage test results are reported in four different ways:

CFM4 is the quantitative volume of air passing through an enclosure under a pressure difference of 4 Pascals (equivalent to a 5mph wind). This is a measurable physical quantity and the value recommended in the FSS to eliminate toxic gas ingress into the SIP spaces.

CFM25 is the quantitative volume of air passing through an enclosure under a pressure difference of 25 Pascals (0.1-in w.g.). FEMA 453 requires Class 1 Safe Havens be pressurized to 25 Pascals. This is a measurable physical property.

CFM50 is the quantitative volume of air passing through an enclosure under a pressure difference of 50 Pascals (0.2-in w.g.).

ACH50 is the quantitative volume of air passing through an enclosure under a pressure difference of 50 Pascals (0.2-in w.g.) normalized by the enclosure volume. ACH is the abbreviation for air changes per hour. It is the number of times the air volume in a building changes per hour at 50 Pascals of pressure.

Table 6 below shows the CFM50 and ACH50 for each of the SIP spaces pre-weatherization and post-weatherization along with the percentage reduction in leakage. SIP Space 1 and SIP Space 5 did not have pre-weatherization results as initially the whole building was tested. Smaller spaces were selected as the whole building was too large to feasibly convert to a SIP space (re-testing of the new SIP spaces was not completed). The control room, SIP space 4, was abandoned in favor of a smaller adjacent room due to high leakage post manual weatherization and the computers that could not be shut down, which prevented the use of AeroBarrier® in the space.

Table 6: Post-weatherization leakage results

| Safe Haven Space | CFM50 | | ACH50 | | % Reduction |
|---|---|---|---|---|---|
| | Pre-Weatherization | Post-Weatherization | Pre-Weatherization | Post-Weatherization | |
| SIP Space 1 | N/A | 1,871 | N/A | 10.7 | N/A |
| SIP Space 2 | 3,623 | 1,560 | 26 | 11.4 | 57% |
| SIP Space 3 | 385 | 135 | 6 | 2.1 | 65% |
| SIP Space 4 | N/A | N/A | N/A | N/A | N/A |
| SIP Space 5 | N/A | 655 | N/A | 19.4 | N/A |

The AeroBarrier® air sealing was prescribed a target. According to the AeroBarrier® simultaneous air leakage calculations, all the spaces met and surpassed this airtightness target; AeroBarrier® sealing was halted at this point. Table 7 shows the CFM50 and ACH50 for each of the SIP spaces as tested by HPBS post AeroBarrier® along with the percentage reduction. The HPBS results were utilized to calculate the amount of positive pressure required to convert the rooms to positive pressure safe havens.

Table 7: Post AeroBarrier® leakage results

| Safe Haven Space | CFM50 | | ACH50 | | % Reduction |
|---|---|---|---|---|---|
| | Post-Weatherization | Post AeroBarrier® | Post-Weatherization | Post AeroBarrier® | |
| SIP Space 1 | 1,871 | 455 | 10.7 | 2.6 | 76% |
| SIP Space 2 | 1,560 | 182 | 11.4 | 1.3 | 88% |
| SIP Space 3 | 135 | 125 | 2.1 | 1.9 | 7% |
| SIP Space 4 | N/A | 173 | N/A | 3.3 | N/A |
| SIP Space 5 | 655 | 138 | 19.4 | 3.1 | 79% |

The AeroBarrier® air sealing methods improved the air barrier integrity for all the proposed SIP spaces between 7% and 88% over the post manual weatherization leakage rates. Air leakage was still observed behind many electrical penetrations including HVAC thermostats, and fire and smoke alarms. These were covered with plastic to protect them from the AeroBarrier® Sealant but had hidden holes behind them.

Figure 1 displays the post AeroBarrier® leakage rate in comparison to the FEMA 453 estimates for air leakage rates based on construction types. FEMA 453 estimates are reported in square feet of floor area at 25 Pascals and do not take into consideration the ceiling height and volume of the space.

Figure 1: Post AeroBarrier® results compared to FEMA 453 estimates

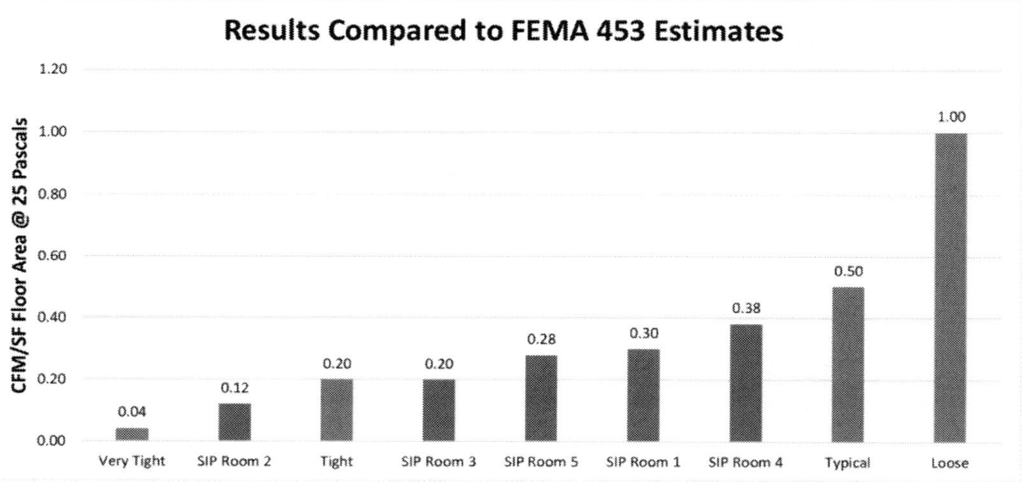


Figure 2 is an estimated leakage rate based on construction type to estimate the filtration and positive pressure system size under the FEMA 453 standard. All the SIP locations were between tight and typical except for SIP Room 2, which was a small prefabricated metal constructed portable building.

Figure 2: Estimated leakage rate per square foot of floor space to achieve 0.1 iwg (25Pa) overpressure

| Construction Type | cfm per square foot of floor area |
|---|---|
| **Very tight:** 26-inch thick concrete walls and roof with no windows | 0.04 |
| **Tight:** 12-inch thick concrete or block walls and roof with tight windows and multiple, sealed penetrations | 0.20 |
| **Typical:** 12-inch thick concrete or block walls with gypsum wall board ceilings or composition roof and multiple, sealed penetrations | 0.50 |
| **Loose:** Wood-frame construction without special sealing measures | 1.00 |

Table 8 shows the combined AeroBarrier®, and manual air sealing methods improved the airtightness for all the proposed SIP spaces between 68% and 95% over the original leakage rates.

Table 8: Air infiltration improvement for SIP spaces

| **Safe Haven Space** | **ACH50** | | | **% Reduction** |
|---|---|---|---|---|
| | **Pre-Weatherization** | **Post-Weatherization** | **Post AeroBarrier®** | |
| SIP Space 1 | N/A | 10.7 | 2.6 | 76% |
| SIP Space 2 | 26.5 | 11.4 | 1.3 | 95% |
| SIP Space 3 | 5.9 | 2.1 | 1.9 | 68% |
| SIP Space 4 | N/A | N/A | 3.3 | N/A |
| SIP Space 5 | N/A | 19.4 | 3.1 | 84% |

Figure 3 compares the ACH50 pre-weatherization, post-weatherization and AeroBarrier® air infiltration results.

Figure 3: Pre-weatherization, post-weatherization and AeroBarrier® comparison

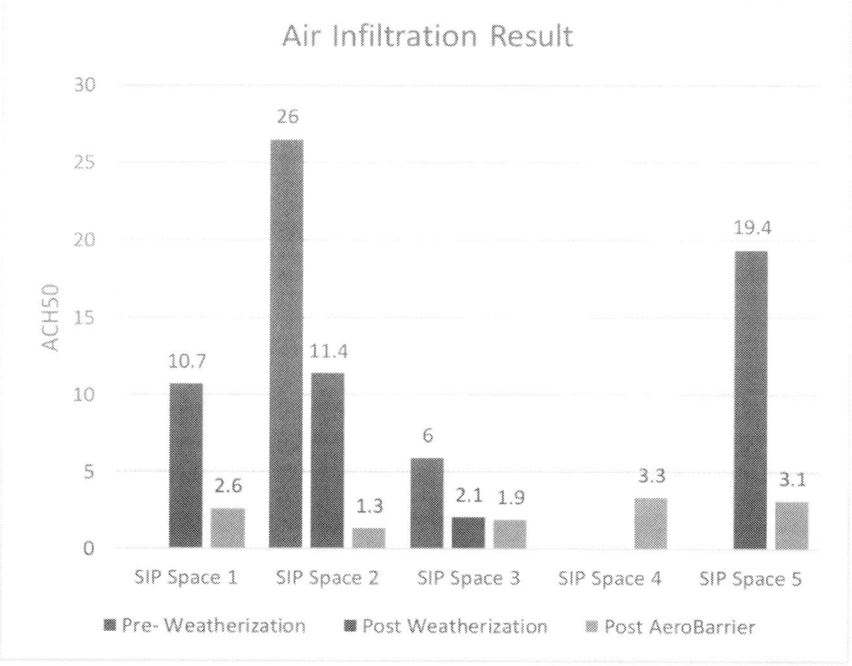


Life Support Systems

MineARC re-calculated the vulnerability for each SIP space based on the post AeroBarrier® leakage rates. All SIP spaces still had a vulnerability of 100% and therefore required the installation of independent life support systems to be deemed zero percent vulnerability safe havens.

Figure 4 shows the toxic gas concentration, duration of the release, and vulnerability for each SIP space. As the largest MCE was a full tank rupture the toxic concentrations were extremely high, but the duration was short at one minute.

Figure 4: SIP space vulnerability post AeroBarrier®

| Safe Haven Location | TOXIC GAS | | | | |
|---|---|---|---|---|---|
| | Worst Case Threat Scenario | Toxic External [ppm] | Duration of Reslease [min] | ACM | Vulnerability |
| SIP Space 1 | Tank Rupture | 585,000 | 1 | 0.032 | 100% |
| SIP Space 2 | Tank Rupture | 652,000 | 1 | 0.042 | 100% |
| SIP Space 3 | Tank Rupture | 660,000 | 1 | 0.007 | 100% |
| SIP Space 4 | Tank Rupture | 412,000 | 1 | 0.042 | 100% |
| SIP Space 5 | Tank Rupture | 600,000 | 1 | 0.028 | 100% |

The leakage rate at 4 Pascals of pressure difference is slightly more than would occur under a 5 mile per hour wind used in the dispersion model in the FSS report based on climatic data for the facility (to maintain positive pressure and neutralize air infiltration). Post-weatherization infiltration rates in cubic feet per minute at 4 Pascals of pressure for each of the SIP spaces is shown in Table 9.

Table 9: Post AeroBarrier® air leakage results at CFM4

| Safe Haven Space | Volume (Cubic Feet) | CFM4 |
|---|---|---|
| SIP Space 1 | 10,481 | 99 |
| SIP Space 2 | 8,210 | 32 |
| SIP Space 3 | 3,917 | 24 |
| SIP Space 4 | 3,110 | 45 |
| SIP Space 5 | 2,637 | 21 |

The air leakage test results provide a mathematical formula that can be used to calculate the amount of breathable compressed air cylinders (437ft$^3$) that must be added to each candidate SIP space to maintain a 4 Pascal positive pressure to overcome a 5mph design wind coincident with a toxic release.

MineARC Systems was contracted to install life support systems to convert the SIP spaces to positive pressure safe havens. The MineARC AirBANK Positive Pressure System provides rapid pressurization, which is activated and maintained using the AirBANK Control via a simple human-machine interface (HMI) touch screen. MineARC's integrated Aura-FX Gas Monitor ensures breathable air automatically remains within acceptable limits. Additionally, the installation of supplementary oxygen and carbon dioxide scrubbing systems in safe havens, are as needed.

Figure 5 shows the number of 437 cubic feet cylinders required for each safe haven location. The minimum breathable air per person is 2.5cfm (70L/min) per person based on MineARC's experience and testing within the refuge chamber industry.

Figure 5: Quantity of cylinders required for 4Pa positive pressure

| | SAFE HAVEN SPECIFICATIONS | | POSITIVE PRESSURE (@ 4Pa) | | | | | | |
|---|---|---|---|---|---|---|---|---|---|
| Safe Haven Location | Occupancy | Duration | CFM | CFM per Floor Space | ACH | No. Cylinders | No. Cylinder Racks | Breathable Air per Person [CFM] | Breathable Air per Person [LPM] |
| Safe Haven Space 1 | 40 | 4 | 99 | 0.10 | 0.6 | 55 | 5 | 2.5 | 70.8 |
| Safe Haven Space 2 | 6 | 4 | 32 | 0.03 | 0.2 | 18 | 2 | 5.4 | 152.9 |
| Safe Haven Space 3 | 5 | 4 | 24 | 0.06 | 0.4 | 14 | 2 | 4.8 | 135.9 |
| Safe Haven Space 4 | 10 | 4 | 45 | 0.14 | 0.9 | 25 | 3 | 4.5 | 127.4 |
| Safe Haven Space 5 | 40 | 4 | 21 | 0.07 | 0.5 | 12 | 2 | 0.6 | 17.0 |

The maximum $CO_2$ concentration is at 1% and $O_2$ between 19.5% and 23.5% based on OSHA limits (refer to Table 2). Due to the volume and number of people in Safe Haven Space 5, it required supplemental oxygen and carbon dioxide scrubbing to maintain the internal gas concentrations within specified limits. Three sodium chlorate oxygen candles and two AirGEN $CO_2$ scrubbers were included in the design.

Figure 6: Internal gas concentrations in safe havens

| | SAFE HAVEN SPECIFICATIONS | | POSITIVE PRESSURE (@ 4Pa) | | | O2 & CO2 CONCENTRATIONS (@ 4Pa) | | | |
|---|---|---|---|---|---|---|---|---|---|
| Safe Haven Location | Occupancy | Duration | CFM | CFM per Floor Space | ACH | Indoor $O_2$ % | No. $O_2$ Candles | Indoor $CO_2$ % | No. AirGEN Scrubbers |
| Safe Haven Space 1 | 40 | 4 | 99 | 0.10 | 0.6 | 20.45 | 0 | 0.50 | 0 |
| Safe Haven Space 2 | 6 | 4 | 32 | 0.03 | 0.2 | 20.84 | 0 | 0.17 | 0 |
| Safe Haven Space 3 | 5 | 4 | 24 | 0.06 | 0.4 | 20.77 | 0 | 0.23 | 0 |
| Safe Haven Space 4 | 10 | 4 | 45 | 0.14 | 0.9 | 20.66 | 0 | 0.31 | 0 |
| Safe Haven Space 5 | 40 | 4 | 21 | 0.07 | 0.5 | 18.67 | 3 | 2.04 | 2 |

In 2018 MineARC completed installation and training on the five safe haven spaces. Each safe haven was tested during the commissioning process to ensure it could maintain acceptable positive pressure.

## CONCLUSIONS

The conversion of shelter-in-place (SIP) locations to zero percent vulnerability safe havens is a four-step process consisting of the following:

1) Identifying the correct SIP location based on ease of sealing and small volume based on the required occupancy.
2) Manual air sealing of SIP spaces.
3) AeroBarrier® sealing of SIP spaces.
4) Installation of life support equipment.

Identifying SIP spaces that can seal from the outset is crucial to a successful result. In the case study, three of the five original locations were abandoned in favor of rooms in adjacent locations due to high leakage rates or reducing the size to make the project more cost effective. Control rooms pose the most difficulty in trying to seal due to the false floor and a myriad of cable tray and pass-throughs installed for the computer systems. The smaller the room and the more robust the construction (i.e. prefabricated metal or concrete buildings), the better it seals.

Manual air sealing needs to be overseen by someone with familiarity in building tightness. While most of the tasks are general construction, standard contractors are not trained to work at the level of precision needed to ensure proper sealing. Even a tiny orifice or ceiling tile not sealed correctly can ruin a leakage test result. HPBS and MineARC Systems have developed a system for ensuring airtightness in construction methodology and by avoiding un-sealable spaces or spaces that are not cost effective to seal.

For future construction projects, incorporation of SIP spaces should occur during the design phase. Air barrier commissioning is standard in new construction and should include the following tasks at a minimum: design review of the plans with specifications for the air barrier and mechanical system - specifically as these relate to spaces designated for use as safe havens; quality assurance inspections during construction to ensure a continuous air barrier is in place; and final air leakage testing of the continuous air barrier to ensure the leakage rate of the safe haven spaces are within acceptable limits.

Complementary life support equipment is essential to sustain life within an airtight environment. For personnel to remain within the sealed safe haven, the environment must be habitable. Air is to be free of toxins such as carbon dioxide and acceptable levels of oxygen maintained. Positive pressure systems reduce the ingress of toxins and provide some or all of the required source of breathable air. Additional equipment such as compressed air, oxygen cylinders and air scrubbers are used to meet the shortfall. Furthermore, internal conditions must consider the human condition, supporting workable body temperature and humidity.

Reducing the risk of harm to personnel is a priority for emergency response planning. While the likelihood of an MCE occurring may be low, the risk to on-site personnel in the event of an incident is high. Converting shelter-in-place (SIP) locations to zero vulnerability safe havens improves emergency response.

## REFERENCES

ASTM E779-10, *Standard Test Method for Determining Air Leakage Rate by Fan Pressurization*, ASTM International, West Conshohocken, PA, 2019, www.astm.org

Brake, Fellow, & Bates (1999). *Criteria for the design of emergency refuge stations for an underground metal mine*. Aus. IMM Proceedings, No.2, 1999, pp 1–7.

Department of Mines and Petroleum, 2013, *Refuge chambers in underground mines — guideline: Resources Safety*, Department of Mines and Petroleum, Western Australia, 41 pp.

Epstein, Y., & Moran D. (2006). *Thermal Comfort and the Heat Stress Indices,* Industrial Health, 44, 388–398.

FEMA 453. (2006). *Design Guidance for Shelters and Safe Rooms*. Federal Emergency Management Agency

Health & Safety Executive UK (2008). *Guidance and Information on the role and design of safe havens in arrangements for escape from mines.* Retrieved from http://www.hse.gov.uk/pubns/mines08.pdf

ITA Working Group 5 (2016). *Provisions for refuge chambers in tunnel projects under construction.* ITA Report N°014 - V2 / March 2018

Mines Safety and Health Administration (2008). *30 CFR Parts 7 and 75 Refuge Alternatives for Underground Coal Mines; Final Rule.*

National Institute for Occupational Safety and Health (2007). *Research Report on Refuge Alternatives for Underground Coal Mines.* Retrieved from http://www.cdc.gov/niosh/docket/archive/pdfs/NIOSH-125/125 ResearchReportonRefugeAlternatives.pdf

Occupational Safety & Health Administration (2013). *Permit-required confined spaces.* Retrieved from http://www.osha.gov/pls/oshaweb/owadisp.show_document?p_table=STANDARDS&p_id=9 797

Suming Du. (2014). *On the effectiveness of shelter-in-place as a measure to reduce harm from atmospheric releases. Journal of Emergency Management, Vol. 12, No, 3, May/June 2014.*

Venter. J. (1998). *Portable Refuge Chambers: Air or Tomb in Underground Escape Strategies.* School for Mechanical and Materials Engineering, Potchefstoom University, Republic of South Africa

Worksafe Australia National Standard (2005). *Australian Standard AS: NZS 1716:2003 Respiratory protective devices.*

Yantosik. G. (2006). *Shelter-in-Place Protective Action Guide*. The Chemical Stockpile Emergency Preparedness Program (CSEPP). U.S. Department of Energy

# 589008 How the Field of Human Factors Helps Solve the Riddle of Serious Injuries and Fatalities

**Abstract:**

Process and occupational safety are often treated as distinct endeavors, with the former tending to focus on facility system function and the latter on individual worker wellbeing. Though other differences certainly exist between the two, both endeavors are generally purposed to prevent incidents that can cost lives, health, time, and money, in addition to employee morale and public trust. The overarching thesis of this paper is that the field of Human Factors is uniquely positioned to effect positive change in both of these major safety categories.

The backdrop for our argument is the recent finding that while injury and illness incidence rates published by the Bureau of Labor statistics have decreased in recent years, the occurrences of serious injuries and fatalities have declined at a significantly slower rate. We discuss possible explanations for this finding but note that regardless of the preferred explanation, the trend should motivate scientists, engineers, and other safety professionals to revisit long-held accident causation theories holding that near misses – these are incidents that could have resulted in serious injuries and fatalities but did not – are precursors of future serious incidents. While this idea still has merit, the just-mentioned statistics warrant reconsideration of the relations between serious and less-serious incidents to explain why the rates of the two are not decreasing in lockstep. We suggest these data indicate additional (human) factors are involved.

To address this riddle and to support our general thesis, we (a) review core tenets of process safety, occupational safety, and the field of Human Factors, (b) point out that the human is always an element common to both process and occupational safety contexts, and (c) discuss how preventing or mitigating human error can benefit both. We offer specific examples from both safety contexts to show how Human Factors perspectives and techniques can help. Examples touch on standards, policies, and other administrative controls (e.g., procedures), human-machine interfaces of both workplace hardware (e.g., valves) and software (e.g., SCADA systems), environments (e.g., physical, regulatory, and cultural), and communications between personnel (e.g., elements of team resource management). We conclude that fundamental Human Factors considerations can help prevent and mitigate human performance gaps and therefore increase the rate of reduction in serious injuries and fatalities in both process and occupational safety.

# 589233 Initial Conditions Used in Atmospheric Dispersion Models:Issues with Commonly Used Assumptions for Gas Jets and Plumes

**Abstract:**

When simulating jets and plumes, all atmospheric dispersion models require initial conditions that appropriately parameterize the behavior of a gas release to the atmosphere. Often, process conditions are such that loss of primary containment of a gas will result in choked flow conditions. A great deal of experimental and theoretical consideration has been given to the flow from stagnation conditions to choked flow conditions, but less experimental work has centered on describing the flow through the depressurization process to atmospheric pressure where atmospheric dispersion models would be applicable. Traditional modeling of the depressurization process has taken the simplified approach of assuming a (depressurized) “top hat” profile (laterally constant) with application of the principles of conservation of mass, momentum, and energy. Such assumptions are known to result in predicted depressurized velocities that exceed the (sonic) velocity at the point of loss of containment, and because of the energy balance, the associated temperatures are significantly less than temperatures at the choked conditions. Physically, these effects could possibly self-cancel because as jet velocity slows, the decrease in kinetic energy of the jet will increase the jet temperature. However, no atmospheric dispersion model used for routine hazard assessment purposes accounts for kinetic energy in the energy balance. The traditional modeling approaches of the depressurization process are reconsidered based on analysis of available data, and an alternative approach to modeling the depressurization process is proposed. Example cases showing the important differences between the two approaches are discussed.

# Fault Tree Uncertainty Analysis

**Raymond "Randy" Freeman**
**S&PP Consulting**
**12303 Lake Shore Ridge**
**Houston, TX 77041**
**Cell: 713 408 0357**
**Rafree@yahoo.com**

**For Presentation at the 2020 Spring National AIChE Meeting**
**Houston, Texas**

**March 30, 2020**

## ABSTRACT

Fault tree analysis (FTA) is a widely used methodology in the process industries. FTA is used for the development of failure mechanisms, computation of failure frequencies and the determination of the probability of failure on demand of safety systems. Much of the data used in a FTA study are uncertain. For example, the failure rate of a pump is often not known with great precision. Likewise the failure rates of instrumentation are often known only within some defined limits. The common practice, used by analysts in the quantification of a fault tree, is to use the most likely or best guess as to the needed failure rate data. The use of best guess values as data inputs to the quantification of a fault tree creates uncertainty in the computed results.

This paper presents a general methodology for the determination of the impact of uncertainty on the results of a fault tree study. The general methodology is based on the mathematics of propagation of error and variance contribution analysis. For a fault tree that contains 10 event minimal cut sets, the methodology will compute the variance in the probability (or frequency) of the top event with a maximum error of 5%. For 5 event minimal cut sets the error in variance of the top event probability (or frequency) will be no more than 2%. An example is presented to illustrate the application of the fault tree uncertainty analysis methodology to a real world problem.

Keywords

Quantitative Risk Assessment, Fault Tree Analysis, Uncertainty Analysis

## 1. Background

Fault tree analysis is a widely used method for representation of failure mechanisms in chemical plants. The methodology is described in the CCPS CPQRA guideline book [Ref. 1] and is extensively discussed in Chapter 9 of Lees [Ref. 2]. Lacking in these discussions is a systematic method to quantify the uncertainty of the results of the fault tree calculations.

Fault tree analysis has also been used in the aircraft and aerospace industries, nuclear power plant review and is recognized by the OSHA Process Safety Management (PSM) standard [Ref. 3] as a method for process hazards analysis.

## 2. Review of Error Propagation and Variance Contribution Analysis (VCA) Methodology

The mean and variance of a function of random variables can be approximated using the method described by Haugen [Ref. 4] and applied by Freeman [Ref. 5, 6, 7]. Define an arbitrary function of a set of random variables, $x_i$, as:

Let

$$Y = F(x_i) \qquad \text{(Eq 1)}$$

The mean of Y can be estimated using the following approximation:

$$E(Y) = F[\, E(x_i)] \qquad \text{(Eq 2)}$$

Where:

$E(Y)$ = expected value of random variable Y = mean of Y

$E(x_i)$ = expected value of random variable $x_i$ = mean of $x_i$

The variance of Y can likewise be estimated as:

$$V(Y) = \sum_{i=1}^{n} \left[ \frac{\partial Y}{\partial x_i} \right]^2 V(x_i) \qquad \text{(Eq 3)}$$

Where:

V(Y) = variance of random variable Y as defined above in Equation 1

$V(x_i)$ = variance of random variable $x_i$ as defined above in Equation 1

Note that the variance is simply the square of the standard deviation. Using the variance will simplify the mathematics that is described below. The contribution of each independent random variable to the overall variance in the function is:

$$V(Y \text{ from } x_i) = \left[\frac{\partial Y}{\partial x_i}\right]^2 V(x_i) \qquad \text{(Eq 4)}$$

The relative contribution of each term to the overall variance V(Y) is a measure of the importance in the uncertainty in the particular random variable, $x_i$. In effect, this is a sensitivity analysis combined with a uncertainty evaluation. The variance contribution combines the sensitivity in the answer to changes in the uncertain random variable, $x_i$, with a measure of the uncertainty in the random variable, $x_i$. The overall variance in Y is found by summing the sensitivity weighted variances from each random variable.

**Numerical Estimate of the Sensitivity**

Let us return to the fundamental definition of the derivative.

As before, set y = F(xi)

$$\frac{\partial y}{\partial x} = \lim_{\Delta xi \to 0} \left[\frac{F(xi+\Delta xi) - F(xi)}{\Delta xi}\right] \qquad \text{(Eq 5)}$$

We are interested in a numerical estimate of the derivative. Therefore, we will make the following approximation.

$$\frac{\partial y}{\partial x} \sim \left[\frac{F(xi+\Delta xi) - F(xi)}{\Delta xi}\right] \qquad \text{(Eq 6)}$$

**Linear Functions**

Let us examine the approximation for a simple linear function.

Let Y = AX + B (Eq 7)

What is the sensitivity of Y to a change in variable X? By simple calculus:

$$\frac{\partial y}{\partial x} = A \qquad \text{(Eq 8)}$$

We can also find the sensitivity using the numerical approximation as:

$$\frac{\partial y}{\partial x} \sim \left[\frac{F(xi+\Delta xi)-F(xi)}{\Delta xi}\right] \qquad \text{(Eq 9)}$$

$$\frac{\partial y}{\partial x} \sim \left[\frac{(A(xi+\Delta xi)+B)-(A(xi)+B)}{\Delta xi}\right] \qquad \text{(Eq 10)}$$

$$\frac{\partial y}{\partial x} \sim \left[\frac{(A(xi+\Delta xi))-(A(xi))}{\Delta xi}\right] \qquad \text{(Eq 11)}$$

$$\frac{\partial y}{\partial x} \sim \left[\frac{((A\,xi+A\,\Delta xi))-(A(xi))}{\Delta xi}\right] \qquad \text{(Eq 12)}$$

$$\frac{\partial y}{\partial x} \sim \left[\frac{A\,\Delta xi}{\Delta xi}\right] \qquad \text{(Eq 13)}$$

$$\frac{\partial y}{\partial x} = A \qquad \text{(Eq 14)}$$

Note that for a simple linear equation the numerical result for the sensitivity is exactly the same as the analytical expression. There is no restriction on the size of the perturbation ($\Delta xi$) used in the calculations. Previously, Freeman [Ref 7] has suggested using a 10% perturbation in xi to do the sensitivity calculations. However any size of perturbation will work and generate the correct sensitivity.

This is a general result. As long as the function can be expressed as a linear function of a number of variables, xi, the sensitivity can be directly calculated by a simple perturbation of the function. The resulting sensitivity will be exactly the same as would be obtained using analytical methods.

**Non-Linear Expressions**

In the quantification of a fault tree, the top event frequency is computed using failure rates of the component devices and time that measures the interval between system validations. In a minimal cut set of two or more basic events, the time variable may be the same. For example the test interval for a valve may be the same for a pressure transmitter. A minimal cut set involving both the valve and the pressure transmitter would create a 2rd order term in the test interval. If the test interval is uncertain, how do we compute the sensitivity of the top event frequency with respect to the test interval? We will now explore how to deal with this type of problem when completing the uncertainty analysis.

As before, let

$$y = F(xi) = Axi^{n} + B \qquad \text{(Eq 15)}$$

What is the sensitivity of Y to a change in variable X? By simple calculus:

$$\frac{\partial y}{\partial x} = nAxi^{n-1} \qquad \text{(Eq 16)}$$

Let us evaluate the numerical approximation for this case. If n=2, the function of interest is:

$$y = F(xi) = Axi^2 + B \quad \text{(Eq 17)}$$

From basic calculus, the exact analytical expression for the sensitivity is:

$$\frac{\partial y}{\partial x} = 2\,A(xi) \quad \text{(Eq 18)}$$

If xi = 1

$$\frac{\partial y}{\partial x} = 2\,A \quad \text{(Eq 19)}$$

Again the approximation of the derivative is:

$$\frac{\partial y}{\partial x} \sim \left[\frac{(A(xi+\Delta xi)^2+B)-(Axi+B)}{\Delta xi}\right] \quad \text{(Eq 20)}$$

$$\frac{\partial y}{\partial x} \sim \left[\frac{F(xi+\Delta xi)-F(xi)}{\Delta xi}\right] \quad \text{(Eq 21)}$$

Inserting the equation for F(xi) we obtain:

$$\frac{\partial y}{\partial x} \sim \left[\frac{(A(xi+\Delta xi)^2+B)-(Axi+B)}{\Delta xi}\right] \quad \text{(Eq 22)}$$

$$\frac{\partial y}{\partial x} \sim \left[\frac{(A\,(xi^2+2\Delta xi+\Delta xi^2)\,+B)-(Axi+B)}{\Delta xi}\right] \quad \text{(Eq 23)}$$

$$\frac{\partial y}{\partial x} \sim \left[\frac{A\,(xi^2+2\Delta xi+\Delta xi^2)\,-Axi}{\Delta xi}\right] \quad \text{(Eq 24)}$$

$$\frac{\partial y}{\partial x} \sim A\left[\frac{(xi^2+2\Delta xi+\Delta xi^2)\,-xi}{\Delta xi}\right] \quad \text{(Eq 25)}$$

Setting xi =1, and using a 10% perturbation as $\Delta xi = 0.1\,xi = 0.1$

$$\frac{\partial y}{\partial x} \sim A\left[\frac{(1^2+(2)(0.1)1+0.1^2\,)\,-1}{0.1\,(1)}\right] \quad \text{(Eq 26)}$$

$$\frac{\partial y}{\partial x} \sim A\left[\frac{(2)(0.1)+0.01}{0.1}\right] \quad \text{(Eq 27)}$$

$$\frac{\partial y}{\partial x} \sim A\left[\frac{0.21}{0.1}\right] \quad \text{(Eq 28)}$$

$$\frac{\partial y}{\partial x} = 2.1\,A \quad \text{(Eq 29)}$$

Note that the numerical estimate (Eq 29) is approximately 5% greater than the analytical value (Eq 19). As shown in Table 1, for higher order equations the error will increase as the size of the perturbation becomes more important. Table 2 shows the impact of reducing the perturbation size to 1%. For $10^{th}$ order variables (say a minimal cut set with 10 basic events) the error is 5%. For almost all engineering evaluations, a maximum error of 5% should be adequate.

For cases where the test frequency, repair time, or mission time appears in a minimal cut set 3 or more times, I suggest that the following perturbation in time be used to compute the sensitivity of the top event:

$$\Delta time = 0.01\ time \qquad \text{(Eq 30)}$$

The sensitivity of the top event frequency to all uncertain variables can be computed using a perturbation of 1%.

$$\Delta xi = 0.01\ xi \qquad \text{(Eq 31)}$$

## 3. Fault Tree Analysis

Fault trees are often used to analyze potential failures of an engineered system. A fault tree is a failure logic diagram that shows the relationship of device or system failures leading to an accident. Fault tree analysis is covered in detail in the CCPS CPQRA book [Ref. 1]. A fault tree is constructed by starting with a top event or condition and asking the question:

What conditions lead to this outcome?

Each condition in the next layer down is then analyzed by asking the same question. What conditions lead to this outcome? The logic of how these intermediate conditions (termed intermediate events) are related is expressed using AND and OR logic gates. An AND gate output is true if and only if (iff) all of the inputs to the AND gate are all true simultaneously. An OR gate output is true iff one of the inputs to the OR gate is true.

Once the fault tree logic structure is created, numerical data can be used to calculate the frequency or probability of the top event of concern occurring. For fault trees where a basic event is found in multiple branches of the tree, such as power failure, the logic structure of the tree must be simplified to remove the impact of the repeated events on the top event of concern. This simplification is done using Boolean Algebra. Each resulting grouping of basic events can cause the top event to occur. These groupings of basic events are called minimal cut sets. As a simple example consider the Boolean Algebra representation of a fault tree :

T = BE11 + BE1 • BE2 + BE1 • BE3 + BE2 • BE9 + BE3 • BE7 (Eq 32)

Where:

T = Fault Tree Top Event

BE11 = Basic Event 11

BE1 = Basic Event 1

BE2 = Basic Event 2

BE3 = Basic Event 3

BE7 = Basic Event 7

BE9 = Basic Event 9

\+ = Boolean Algebra Addition Symbol

• = Boolean Algebra Multiplication Symbol

For this example the top event T, the minimal cut sets are:

BE11

BE1 BE2

BE1 BE3

BE2 BE9

BE3 BE7

For example the existence of basic events BE1 and BE3 guarantees that the top event T will occur. Publically available standard computer software such as SAPHIRE [Ref. 8] may be used to find the minimal cut sets of a large fault tree.

The representation of the top event in equation 32 can now be used to compute the probability or frequency of occurrence of the top event. If the devices describe by the basic events are non-repairable and are tested after a period of time TI, the probability of top event, T, may be calculated as:

$$\text{Prob (T)} = \lambda_{11}\,TI/2 + (\lambda_1\,TI/2)\,(\lambda_2\,TI/2) + (\lambda_1\,TI/2)\,(\lambda_3\,TI/2) + (\lambda_2\,TI/2)\,(\lambda_9\,TI/2) + (\lambda_3\,TI/2)\,(\lambda_7\,TI/2) \qquad \text{(Eq 33)}$$

Where:

$\lambda_{11}$ = failure rate of device 11 in basic event BE11, failures per unit time

$\lambda_1$ = failure rate of device 1 in basic event BE1, failures per unit time

$\lambda_2$ = failure rate of device 2 in basic event BE5, failures per unit time

$\lambda_3$ = failure rate of device 3 in basic event BE3, failures per unit time

$\lambda_7$ = failure rate of device 7 in basic event BE7, failures per unit time

$\lambda_9$ = failure rate of device 9 in basic event BE9, failures per unit time

TI = test interval, time

Equation 33 assumes that the device failure rates are small and the product $\lambda$ TI $< 0.1$. The other major assumption is that the values of all of the failure rates and test intervals are known. In reality the failure rate data are not known exactly with only a range or probability distribution representing the state of knowledge.

## 4. Example Problem

The following example is taken from the ISA technical report on fault tree analysis [Ref. 9]. This example should not be considered as a recommendation by either ISA or AIChE. The example is presented to allow for the demonstration of the methods that can be used in a fault tree study to quantify the uncertainty in the resulting calculated failure frequency or failure probability.

Consider the interlock on the intermediate storage tank T101 shown on Piping and Instrument Diagram in Figure 1. The interlock block diagram is shown in Figure 2. The interlock is intended to prevent an abnormal condition in the tank which could lead to a uncontrolled release of the tank contents. The process hazards analysis team has recommended that the interlock be designed to safety integrity level 2 (SIL 2) with a target probability of failure on demand (PFD) of no greater than 1E-2 or a risk reduction factor (RRF) of 100. Does the proposed interlock shown on Figure 1 meet the SIL 2 target? What is the uncertainty in the predicted PFD of the interlock?

A fault tree for the failure of this interlock is shown in Figure 3. A minimal cut set analysis has been completed and the resulting minimal cut sets are presented in Table 3. We can now write the Boolean Algebra equation that represents this fault tree as:

$$T = PE + TS1 \bullet TS2 + LS1 \bullet LS2 + FT1 \bullet FT2 + FT2 \bullet FT3 + FT1 \bullet FT3 + BV1 \bullet BV2 + BV1 \bullet SOL1 + BV2 \bullet SOL2 + SOL1 \bullet SOL2 + PT1 \bullet PT2 \quad \text{(Eq 34)}$$

Assuming that the interlock is non-repairable until tested at time TI and using lamda-time ($\lambda T$) approximation of the failure rate, the Boolean representation of the fault tree may now be converted to a failure probability model as:

$$\begin{aligned} Prob(T) = {} & Prob(PE) + (\lambda_{TS1}\,TI/2)(\lambda_{TS2}\,TI/2) + (\lambda_{LS1}\,TI/2)(\lambda_{LS2}\,TI/2) + (\lambda_{FT1}\,TI/2)(\lambda_{FT2}\,TI/2) + \\ & (\lambda_{FT2}\,TI/2)(\lambda_{FT3}TI/2) + (\lambda_{FT1}\,TI/2)(\lambda_{FT3}\,TI/2) + (\lambda_{BV1}\,TI/2)(\lambda_{BV2}\,TI/2) + \\ & (\lambda_{BV1}\,TI/2)(\lambda_{SOL1}TI/2) + (\lambda_{BV2}\,TI/2)(\lambda_{SOL2}\,TI/2) + \\ & (\lambda_{SOL1}\,TI/2)(\lambda_{SOL2}\,TI/2) + (\lambda_{PT1}\,TI/2)(\lambda_{PT2}\,TI/2) \quad \text{(Eq 35)} \end{aligned}$$

Where:

Prob(PE) = failure probability of electronic logic solver, assumed to be constant at 5E-3.

$\lambda_{TS1}$ = failure rate of temperature switch 1, failures per unit time

$\lambda_{TS2}$ = failure rate of temperature switch 2, failures per unit time

$\lambda_{LS1}$ = failure rate of level switch 1, failures per unit time

$\lambda_{LS2}$ = failure rate of level switch 2, failures per unit time

$\lambda_{FT1}$ = failure rate of flow transmitter 1, failures per unit time

$\lambda_{FT2}$ = failure rate of flow transmitter 2, failures per unit time

$\lambda_{FT3}$ = failure rate of flow transmitter 3, failures per unit time

$\lambda_{BV1}$ = failure rate of block valve 1, failures per unit time

$\lambda_{BV2}$ = failure rate of block valve 2, failures per unit time

$\lambda_{SOL1}$ = failure rate of solenoid valve 1, failures per unit time

$\lambda_{SOL2}$ = failure rate of solenoid valve 2, failures per unit time

$\lambda_{PT1}$ = failure rate of pressure transmitter 1, failures per unit time

$\lambda_{PT2}$ = failure rate of pressure transmitter 2, failures per unit time

TI = test interval, time

The electronic logic solver PE is assumed to have a constant failure probability of 5E-3.

Failure rate data for the other devices are present in Table 4. The mode (most likely value) of the failure rate is set equal to the point value taken from the ISA example [Ref. 9]. The triangular probability distribution (Appendix A) was assumed to represent the failure rate data for the equipment items. The upper and lower limits are based on the failure rate data of Appendix 4 of Smith [Ref. 10].

Using the device failure rates represented by the Table 4 column labeled as the mode and the minimal cut sets presented in Table 3, the ISA technical report [Ref.9, page 35] computes an interlock PFD as 7.5E-3 (RRF of 133). This would satisfy the requirement for a SIL 2 interlock {PFD of 1E-2 or a RRF of 100}. Now what is the uncertainty in the interlock PFD?

The first thing to remember is that the PFD should be calculated based on the mean value of the failure rate lambda ($\lambda$), not the mode of the failure rate. For a data set, the mean is the best single point representation of the data set. Table 5 presents the PFD of each minimal cut set using the mean of the device failure rate lambda ($\lambda$). The total of the PFD becomes 1.47E-2 (RRF of 68). This would not satisfy a SIL 2 interlock requirement.

Now we will compute the variance of the predicted probability of top event, T. To compute the variance of the probability of the top event, T, we will need to compute the sensitivity of the top event to each of the variables.

**Sensitivity Using Numerical Perturbation Calculation**

We will use the ISA interlock example previously analyzed by Freeman (Ref. 7). Using public data sources Freeman found that the probability of failure on demand (PFD) was 1.49E-2. To use the perturbation method for the evaluation of the sensitivities requires evaluating the impact of small changes in the uncertain input parameters. To enable these calculations to be completed numbers using 8 significant digits were used. The calculations were completed using Microsoft Excel which uses double-precision floating-point arithmetic compliant with IEEE 754 specification. Excel nominally carries 15 significant figures in calculations. We will complete the calculations with Excel displaying 5 significant digits and round the result at the end of the calculations to 4 significant digits. We will illustrate the calculations by looking at the Flow Transmitter, FT1.

a. First we recalculate the PFD of the example and find it to be 1.4594E-02. This is the base number that we will use in the perturbation calculations.
b. We now make a small change in mean failure rate of FT1. The mean failure rate is 8E-6 failures/hour. A 1% perturbation is used or an increase of 0.08E-6 failures/hour.

c. The perturbed value of the failure rate of FT1 becomes
(8 + 0.08) * 1E-6 failures/hour or 8.08E-6 failures/hour.

d. The perturbed value of the failure rate of FT1 is now used to re-calculate the PFD of the interlock. The resulting re-calculation finds that the PFD is 1.4618E-02

e. The change in the PFD from the base number due to the perturbation of the failure rate of FT1 is now computed as:

delta PFD = 1.4618E-02 - 1.4594E-02 = 2.4556E-05

f. The sensitivity of the PFD to a change in the failure rate of FT1 is now computed by dividing the change in the PFD by the change FT1 failure rate as:

sensitivity = (2.4556E-05)/(0.08E-6 failures/hour) = 307.0 hours
Result is rounded to 4 significant digits.

We now note that in the paper by Freeman (Ref. 7) that equation 32 presents the sensitivity for FT1 as:

$$\frac{\partial Prob(T)}{\partial \lambda FT1} = 307.0 \text{ hr} \qquad \text{(Eq 32 of Ref 7)}$$

The answers are exactly the same. The sensitivity calculations for all of the devices are presented in Table 6. Note that in every case the sensitivity found using the analytical expression and the numerical estimate are the same. This is a general result which will be true for all basic event equipment failure rates found in a fault tree minimal cut set.

Once found, the sensitivities can now be used to compute the variance contribution of each particular device to the overall variance of the top event PFD in the same manner as Freeman previously presented.

The sensitivities of the top event probability are summarized in Table 6. We may now compute the variance contribution of each uncertain variable using Equations 1 – 4 in the methodology section of this paper. The relative variance contribution is expressed as a percent of the total variance of the top event probability. The top event variance is found to be 8.24E-6 (failures per hour)$^2$ or equivalently the top event standard deviation is 2.78E-3 failures per hour.

In this example the variance (measure of uncertainty) is dominated by the flow transmitters. The flow transmitters account for almost 2/3 of the uncertainty in the top event probability. Failures of the flow transmitters appear in 3 of the minimal cut sets. Since the uncertainty in the transmitter failure rates is large, the flow transmitters should contribute a significant amount to the variance of the fault tree top event.

We can use the variance in the top event to define the likelihood of achieving SIL 2 performance of the interlock. The calculations are done using the normal probability distribution. The normal distribution is tabulated as the standard normal distribution using a normalization factor Z.

The standard normal factor, Z, [Ref. 11] is defined as:

$$Z = \left[\frac{xi - E(x)}{\sigma}\right] \qquad \text{(Eq 36)}$$

Where:

$\sigma$ = standard deviation. Note that the variance of a random variable is the square of the standard deviation of the random variable.

E(x) = Expected value of the random variable $x_i$

For the example, the expected value of the probability of the top event of concern was calculated as

E[Prob(T)] =1.47E-2 (Eq 37)

From Table 4, the standard deviation in the probability of the top event is

$\sigma_T$ = 2.78E-3 (Eq 38)

The target PFD for a SIL 2 interlock must not be worse than 1E-2 (RRF of 100).

We compute the normal distribution Z factor at the SIL 2 target value as

Z = (0.01 – 0.0147)/2.78E-3 (Eq 39)

Z = -1.69065 (Eq 40)

Tabulations of standard normal distribution are presented in most statistics books such as Meyer [Ref. 11]. For this value of Z, the corresponding probability that the interlock will achieve SIL 2 performance is 4.55 percent. We can also compute the risk reduction that the proposed interlock design will achieve.

What is the interlock probability of failure on demand (PFD) that we can be 95% certain that the interlock will provide? At the 95% level, the corresponding Z factor is 1.65 [Ref 11]. We now compute the corresponding PFD of the interlock that corresponds to this Z factor.

$$Z = \left[\frac{x95\% - E[Prob(T)]}{\sigma}\right] \quad \text{(Eq 41)}$$

In this case:

$x_{95\%}$ = the PFD that we can be confident that the interlock will achieve

$\sigma_T$ =2.78E-3 (Eq 42)

E[Prob(T)] =1.47E-2 (Eq 43)

Z= 1.65 (Eq 44)

Rearranging equation 40 to find $x_{95\%}$we obtain:

$x_{95\%}$ = E[Prob(T)] + Z $\sigma_T$ (Eq 45)

$x_{95\%}$ = 1.47E-2 + (1.65) (2.78E-3) = 0.0193 (Eq 46)

or a 95% certain risk reduction factor of

RRF = 1/PFD = 1/0.0193 = 52 (Eq 47)

By the same logic there is 5% chance that the risk reduction is a PFD of 0.01 or an RRF of 99. In summary, this interlock is likely to perform as a mid-range SIL 1 interlock, not the SIL 2 interlock desired by process hazards analysis team.

## 5. Methodology for Fault Tree Uncertainty Analysis

We can now generalize the methods used in the above example to a proposed method for the evaluation of the uncertainty in a fault tree. The following step-by-step procedure is modified from that that previously published by Freeman (Ref 7) to indicate the use of numerical methods to compute the sensitivity. Step 8 shown in *italics* indicates the modification..

1. Create the fault tree using standard methods outline in the CPQRA book [Ref 1].
2. Determine minimal cut sets by hand or using standard computer software such as SAPHIRE [Ref 12].
3. Define the needed failure rate data for each basic event.
4. Define those basic events that are considered to be uncertain.
5. For the basic events that are uncertain, define the probability distribution and associated parameters needed to numerically define the probability distribution. Appendix B of this paper presents the needed description of 4 commonly used probability distributions. In many cases the triangular distribution will be selected. The minimum, maximum and mode (most likely) parameters will be needed for the triangular distribution.
6. Compute the mean of each of the uncertain variable
7. Compute the probability or frequency of the fault tree top event of interest using the mean value for each uncertain variable.
8. *Compute the sensitivity of the top event probability or frequency using the numerical methods outlined in this paper.*
9. Compute the variance of each of the uncertain variables
10. Compute the variance contribution of for each of the uncertain variables to the top event probability or frequency using equation 6.
11. Compute the total variance of the top event of interest probability or frequency by summing all of the contributions determined in Step 10.
12. Compute the variance contribution percent of each uncertain variable by dividing the variable contribution (Step 10) by the total variance of the top event of interest (Step 11).
13. Define the level of risk the project is willing to take. What chance will the project management accept for potential failure of the interlock to achieve the desired risk reduction? In the above example, I have used a 5% risk of failure or a 95% certainty that the interlock will achieve a risk reduction factor of 52. In the example, there is a very low probability that the interlock will achieve a SIL 2 PFD of 0.01 or an RRF of 100.

Alternately you can report the 90% range for the top event of interest. For the example the 90% range starts at the 5% RRF of 99 and there is a 95% certainty that that an RRF of at least 52 will be achieved. The expected risk reduction for this interlock is an RRF of 68.

## 6. Conclusions

A general procedure has been presented to quantify the uncertainty in calculations of failure frequencies using the fault tree methodology. The uncertainty analysis procedure is based on the application of propagation of error and variance contribution analysis techniques to the minimal cut sets created during a fault tree study. The procedure is simple and can be incorporated into standard fault tree analysis programs such as SAPHIRE [Ref 8]. This procedure is based on perturbation calculation of the frequency of the top event from changes in the uncertain variables. Incorporation of a numerical perturbation method into standard fault tree software such as SAPHIRE would be easy to implement and would provide the means for evaluation of the uncertainty in the quantified fault tree results.

For the example interlock presented in this paper, the uncertainty in the device failure rates degrades the predicted interlock SIL from meeting SIL 2 requirements to becoming a mid range SIL 1 interlock. This is a general result. Uncertainty in design parameters will reduce the likelihood that the equipment will achieve desired results. The design engineer has two choices:

- Accept the conclusion that the interlock will not perform as SIL 2
- Re-design the interlock using more reliable components to achieve the SIL 2 target

The variance contribution analysis shown in Table 7 indicates that the major contributor to the uncertainty in the interlock performance is the Flow Transmitters. Flow transmitters certified to SIL-2 or SIL-3 performance with lower failure rates could improve the calculated PFD of the interlock.

## 7. References


1. Center for Chemical Process Safety , *Guidelines for Chemical Process Quantitative Risk Analysis – 2rd edition*, American Institute of Chemical Engineers, New York, 2000.
2. Frank P Lees, Loss Prevention in the Process Industries – 2rd Edition, Butterworth, ISBN 0-7506-1547-8, Oxford, UK, 1996
3. US Department of Labor, Occupational Safety and Health Administration, CFR 1910.119 - Process Safety Management of Highly Hazardous Chemicals, 1910.119(e)(2)(vi).
4. Edward B. Haugen, *Probabilistic Approaches to Design*, John Wiley, New York, 1968
5. Raymond. Freeman and Angela Summers, "Evaluation of Uncertainty in Safety Integrity Level Calculations," *Process Safety Progress,* Published online in Wiley Online Library (wileyonlinelibrary.com) DOI 10.1002/prs. 11805, 2016
6. Raymond. Freeman, "General Method for Uncertainty Evaluation of Safety Integrity Level Calculations – Part 2 Analytical Methods," *Process Safety Process*, Published online in Wiley Online Library (wileyonlinelibrary.com) DOI 10.1002/prs 111915, October 2017
7. Raymond. Freeman ,"Error Propagation and Uncertainty Analysis: Application to Fault Tree Analysis", Process Safety Progress, Published Wiley Online Library (wileyonlinelibrary.com/journal/prs) DOI: https://doi.org/10.1002/prs.12080, June 29, 2019

8. US Nuclear Regulatory Commission, "Systems Analysis Programs for Hands-on Integrated Reliability Evaluations (SAPHIRE) – Version 8," NUREG/CR-7039, Vol 3, June 2011.
9. ISA, "Safety Instrumented Functions (SIF) – Safety Integrity Level (SIL) Evaluation Techniques Part 3: Determining the SIL of a SIF via Fault Tree Analysis, ISA Technical Report ISA-TR-84.00.02-2002 – Part 3, Approved 17 June 2002.
10. David J. Smith, *Reliability Maintainability and Risk – 8th Edition*, ISBN 978-0-08-096902-2, Elsevier, London, 2011
11. Paul L. Meyer, Introductory Probability and Statistical Applications, John Wiley, New York, 1972

**Table 1. Comparison of Analytical Sensitivity with Numerical Approximation with Perturbation of 10% in xi for the equation: $y = F(xi) = Axi^n + B$**

| Order of Equation n | Equation for Analytical Sensitivity $\frac{\partial y}{\partial x} = nAxi^{n-1}$ | Sensitivity with xi = 1 and A=1 | Numerical Sensitivity with $\Delta xi = 0.1\,xi$ | Error % |
|---|---|---|---|---|
| 1 | $\frac{\partial y}{\partial x} = 1A(xi)^0$ | 1 | 1 | 0 |
| 2 | $\frac{\partial y}{\partial x} = 2A(xi)^1$ | 2 | 2.1 | 5 |
| 3 | $\frac{\partial y}{\partial x} = 3A(xi)^2$ | 3 | 3.31 | -10% |
| 4 | $\frac{\partial y}{\partial x} = 4A(xi)^3$ | 4 | 4.64 | -16% |
| 5 | $\frac{\partial y}{\partial x} = 5A(xi)^4$ | 5 | 6.11 | -22% |
| 6 | $\frac{\partial y}{\partial x} = 6A(xi)^5$ | 6 | 7.72 | -29% |
| 7 | $\frac{\partial y}{\partial x} = 7A(xi)^6$ | 7 | 9.49 | -36% |
| 8 | $\frac{\partial y}{\partial x} = 8A(xi)^7$ | 8 | 11.44 | -43% |
| 9 | $\frac{\partial y}{\partial x} = 9A(xi)^8$ | 9 | 13.58 | -51% |
| 10 | $\frac{\partial y}{\partial x} = 10A(xi)^9$ | 10 | 15.94 | -59% |

Error = (Analytical- Numerical)/Analytical

**Table 2. Comparison of Analytical Sensitivity with Numerical Approximation with Perturbation of 1% in xi for the equation: $y = F(xi) = Axi^n + B$**

| Order of Equation n | Equation for Analytical Sensitivity $\frac{\partial y}{\partial x} = nAxi^{n-1}$ | Sensitivity with xi = 1 and A=1 | Numerical Sensitivity with $\Delta xi = 0.1\ xi$ | Error % |
|---|---|---|---|---|
| 1 | $\frac{\partial y}{\partial x} = 1A(xi)^0$ | 1 | 1 | 0 |
| 2 | $\frac{\partial y}{\partial x} = 2A(xi)^1$ | 2 | 2.01 | -1% |
| 3 | $\frac{\partial y}{\partial x} = 3A(xi)^2$ | 3 | 3.03 | -1% |
| 4 | $\frac{\partial y}{\partial x} = 4A(xi)^3$ | 4 | 4.06 | -2% |
| 5 | $\frac{\partial y}{\partial x} = 5A(xi)^4$ | 5 | 5.10 | -2% |
| 6 | $\frac{\partial y}{\partial x} = 6A(xi)^5$ | 6 | 6.15 | -3% |
| 7 | $\frac{\partial y}{\partial x} = 7A(xi)^6$ | 7 | 7.21 | -3% |
| 8 | $\frac{\partial y}{\partial x} = 8A(xi)^7$ | 8 | 8.29 | -4% |
| 9 | $\frac{\partial y}{\partial x} = 9A(xi)^8$ | 9 | 9.37 | -4% |
| 10 | $\frac{\partial y}{\partial x} = 10A(xi)^9$ | 10 | 10.46 | -5% |

Error = (Analytical- Numerical)/Analytical

**Table 3. Basic Events in Minimal Cut Sets For Example**

| Cut Set No | BE 1 | BE 2 |
|---|---|---|
| 1 | PE | - |
| 2 | TS1 | TS2 |
| 3 | LS1 | LS2 |
| 4* | FT1 | FT2 |
| 5* | FT2 | FT3 |
| 6* | FT1 | FT3 |
| 7* | BV1 | BV2 |
| 8* | BV1 | SOL1 |
| 9* | BV2 | SOL2 |
| 10* | SOL1 | SOL2 |
| 11 | PT1 | PT2 |

* Minimal cut Set with same basic event as another cut set

**Table 4. Lamda ( λ ) Failure Rate Data With Uncertainty Limits**

| Device | Minimum Failures per 1E6 Hours | Mode Failures per 1E6 Hours | Maximum Failures per 1E6 Hours | Mean Failures per 1E6 Hours | Lamda Variance (Failures per 1E6 Hours)$^2$ |
|---|---|---|---|---|---|
| Flow Transmitters FT1, FT2 and FT3 | 1.0 | 2.9 | 20.0 | 8.0 | 18.3 |
| Pressure Transmitters PT1 and PT2 | 1.0 | 2.3 | 20.0 | 7.8 | 18.8 |
| Temperature Switch TS1 and TS2 | 3.0 | 7.6 | 20.0 | 10.2 | 12.9 |
| Level Switch LS1 and LS2 | 2.0 | 4.6 | 20.0 | 8.9 | 15.8 |
| Block Valves BV1 and BV2 | 0.2 | 2.3 | 10.0 | 4.2 | 4.4 |
| Solenoid Valves SOL1 and SOL2 | 1.0 | 2.3 | 8.0 | 3.8 | 2.3 |

**Table 5. Probability of Minimal Cut Sets Based on Mean Failure Rates**

| Cut Set No | BE 1 | BE 2 | PFD based on mean λ |
|---|---|---|---|
| 1 | PE | - | 5.00E-03 |
| 2 | TS1 | TS2 | 2.00E-03 |
| 3 | LS1 | LS2 | 1.52E-03 |
| 4 | FT1 | FT2 | 1.23E-03 |
| 5 | FT2 | FT3 | 1.23E-03 |
| 6 | FT1 | FT3 | 1.23E-03 |
| 7 | BV1 | BV2 | 3.38E-04 |
| 8 | BV1 | SOL1 | 3.38E-04 |
| 9 | BV2 | SOL2 | 3.38E-04 |
| 10 | SOL1 | SOL2 | 2.77E-04 |
| 11 | PT1 | PT2 | 1.17E-03 |

Table 6. Sensitivity of Example System PFD to Basic Event Failure Rates Based on a 1% Perturbation

| Basic Event Label | Device Type | Mean Failures per 1E6 Hours | Failure Perturbation of 1% of Mean Failures per 1E6 Hours | Perturbed Failure Rate Failures per 1E6 Hours | Total System PFD Using Perturbed Failure Rate of Basic Event | Change in Total System PFD | Sensitivity of PFD to Change in Failure Rate Hours |
|---|---|---|---|---|---|---|---|
| FT1 | Flow Transmitter | 8 | 0.08 | 8.08 | 1.461848E-02 | 2.455603E-05 | 307.0 |
| FT2 | Flow Transmitter | 8 | 0.08 | 8.08 | 1.461848E-02 | 2.455603E-05 | 307.0 |
| FT3 | Flow Transmitter | 8 | 0.08 | 8.08 | 1.461848E-02 | 2.455603E-05 | 307.0 |
| PT1 | Pressure Transmitter | 7.8 | 0.078 | 7.878 | 1.460560E-02 | 1.167179E-05 | 149.6 |
| PT2 | Pressure Transmitter | 7.8 | 0.078 | 7.878 | 1.460560E-02 | 1.167179E-05 | 149.6 |
| TS1 | Temperature Switch | 10.2 | 0.102 | 10.302 | 1.461389E-02 | 1.995945E-05 | 195.7 |
| TS2 | Temperature Switch | 10.2 | 0.102 | 10.302 | 1.461389E-02 | 1.995945E-05 | 195.7 |
| LS1 | Level Switch | 8.9 | 0.089 | 8.989 | 1.460912E-02 | 1.519596E-05 | 170.7 |
| LS2 | Level Switch | 8.9 | 0.089 | 8.989 | 1.460912E-02 | 1.519596E-05 | 170.7 |
| BV1 | Block Valves | 4.2 | 0.042 | 4.242 | 1.460037E-02 | 6.445958E-06 | 153.5 |
| BV2 | Block Valves | 4.2 | 0.042 | 4.242 | 1.460037E-02 | 6.445958E-06 | 153.5 |
| SOL1 | Solenoid Valves | 3.8 | 0.038 | 3.838 | 1.459976E-02 | 5.832058E-06 | 153.5 |
| SOL2 | Solenoid Valves | 3.8 | 0.038 | 3.838 | 1.459976E-02 | 5.832058E-06 | 153.5 |

Table 7. Variance of Top Event Probability Due to Uncertain Device Failure Rates

| Device No | Device Type | Label | Sensitivity of Top Event Probability to Device Failure Rate Hour$^{-1}$ | Variance of Device Failure Rate (Failures per 1E6 Hour)$^2$ | Variance Contribution to Top Event Variance | Contribution to Top Event Variance % |
|---|---|---|---|---|---|---|
| 1 | Logic Solver | PE | - | 0 | 0.00E+00 | 0% |
| 2 | Temperature Switch | TS1 | 195.7 | 12.9 | 4.94E-07 | 6% |
| 3 | Temperature Switch | TS2 | 195.7 | 12.9 | 4.94E-07 | 6% |
| 4 | Level Switch | LS1 | 170.7 | 15.8 | 4.60E-07 | 6% |
| 5 | Level Switch | LS2 | 170.7 | 15.8 | 4.60E-07 | 6% |
| 6 | Flow Transmitter | FT1 | 307 | 18.3 | 1.72E-06 | 21% |
| 7 | Flow Transmitter | FT2 | 307 | 18.3 | 1.72E-06 | 21% |
| 8 | Flow Transmitter | FT3 | 307 | 18.3 | 1.72E-06 | 21% |
| 9 | Block Valve | BV1 | 153.5 | 4.4 | 1.04E-07 | 1% |
| 10 | Block Valve | BV2 | 153.5 | 4.4 | 1.04E-07 | 1% |
| 11 | Pressure Transmitter | PT1 | 149.6 | 18.8 | 4.21E-07 | 5% |
| 12 | Pressure Transmitter | PT2 | 149.6 | 18.8 | 4.21E-07 | 5% |
| 13 | Solenoid valve | SOL1 | 153.5 | 2.3 | 5.42E-08 | 1% |
| 14 | Solenoid valve | SOL2 | 153.5 | 2.3 | 5.42E-08 | 1% |

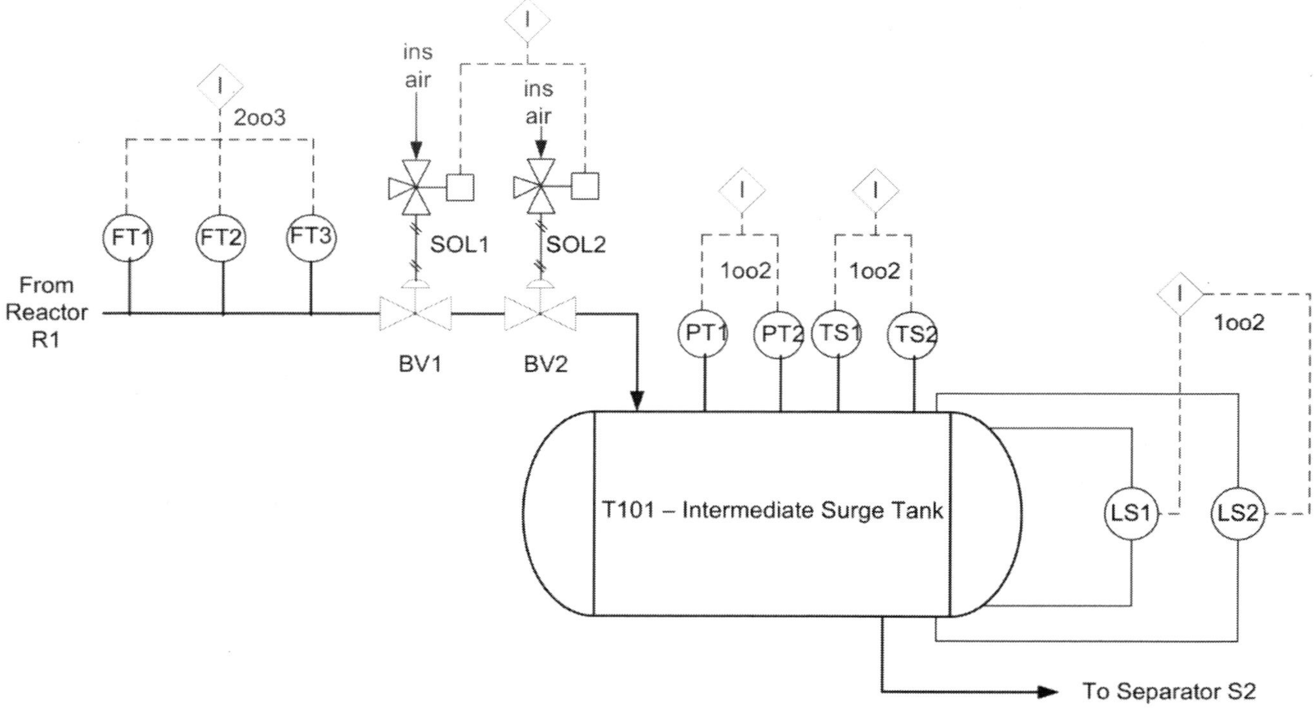


**Figure 1. P&ID Diagram for Example**

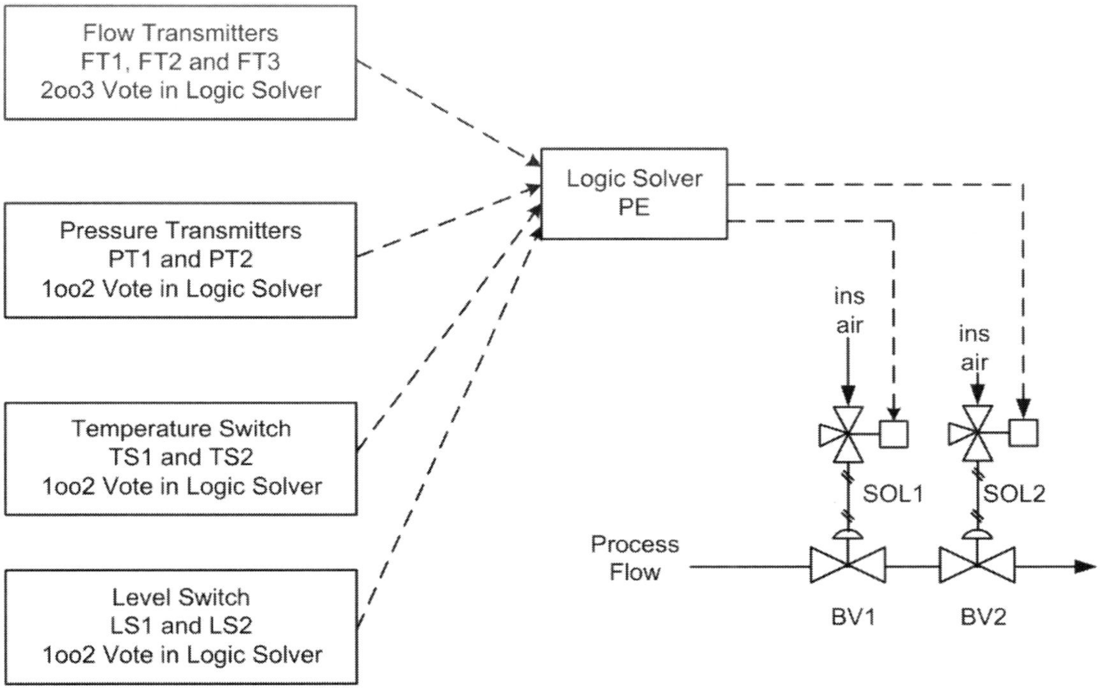


**Figure 2. Interlock Block Diagram**

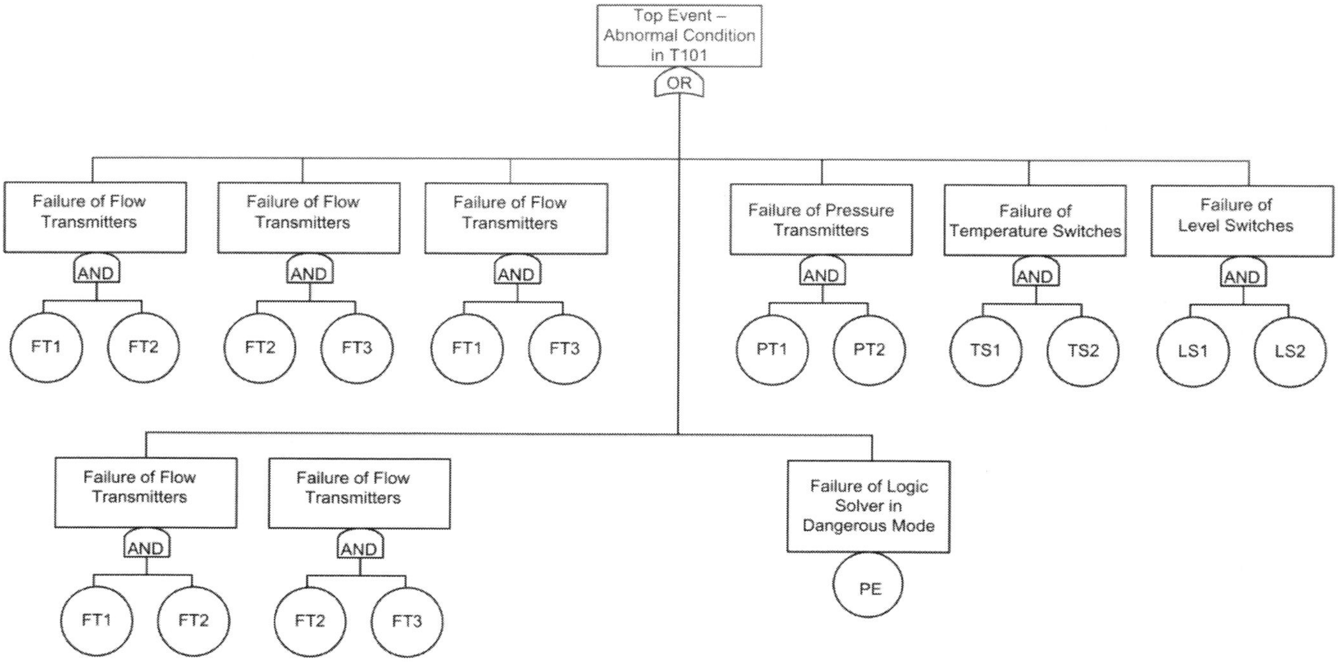


**Figure 3. Fault Tree for Example Interlock**

## Appendix A – Commonly Encounter Probability Distributions

There are four probability distributions used to represent the failure rate of equipment items. These are:

1. Uniform Distribution
2. Triangular Distribution
3. Normal Distribution
4. Log-Normal Distribution

This appendix reviews the properties of these four probability distributions.

### 1. Uniform Distribution

The uniform distribution [Ref. A-1, pg 74; A-2, pg 687-688] is used when a variable is only known within a defined range. Figure A-1 presents a plot of the probability density function (pdf) for the uniform distribution. The distribution is defined using two parameters, A and B.

Where:

A = minimum value of the random variable

B = maximum value of the random variable

The mean or expected value of a random variable, x, that follows the uniform distribution is:

Mean = Expected Value – $E(x) = (A+B)/2$ (Eq A-1)

The variance of a random variable, x, that follows the uniform distribution is:

Variance = $V(x) = (B-A)^2/12$ (Eq A-2)

The standard deviation of a random variable, x, that follows the uniform distribution is:

Standard Deviation = $(\text{Variance})^{1/2} = (B-A)/(12)^{1/2}$ (Eq A-3)

### 2. Triangular Distribution

The triangular distribution [Ref. A-2, pg 686-687] is used when a variable is known to lie within a defined range and a "best guess" or estimate can be made as to the most likely value to the variable. Figure A-2 presents a plot of the probability density function (pdf) for the triangular distribution. The distribution is defined using three parameters, A. B and C.

Where:

A = minimum value of the random variable

B = maximum value of the random variable

C = most likely (mode) of the random variable

The mean or expected value of a random variable, x, that follows the triangular distribution is:

Mean = Expected Value – $E(x) = (A+B+C)/3$ (Eq A-4)

The variance of a random variable, x, that follows the triangular distribution is:

Variance = $V(x) = [A^2 + B^2 + C^2 - AB - AC - BC]/18$ (Eq A-5)

The standard deviation of a random variable, x, that follows the triangular distribution is:

Standard Deviation = $(\text{Variance})^{1/2} = \{[A^2 + B^2 + C^2 - AB - AC - BC]/18\}^{1/2}$ (Eq A-6)

### 3. Normal distribution

The normal distribution [Ref A-2, pg 665-666] is used when a variable when data analysis finds the normal distribution is the best model to describe the spread in the measured variable. Figure A-3 presents a plot of the probability density function (pdf) for the normal distribution. The distribution is defined using two parameters, $\mu$ and $\sigma$.

Where:

$\mu$ = mean of the measured data for variable

$\sigma$ = standard deviation of the measured data for variable

The expected value of a random variable, x, that follows the normal distribution is:

Expected value = mean = $E(x) = \mu$ (Eq A-7)

The variance of a random variable, x, that follows the normal distribution is:

Variance = square of standard deviation = $V(x) = \sigma^2$ (Eq A-8)

The standard deviation of a random variable, x, that follows the normal distribution is:

Standard deviation = $\sigma$ (Eq A-9)

### 4. Lognormal Distribution

The lognormal distribution [Ref. A- 2, pg 658-659] is used when a variable when data analysis finds the lognormal distribution is the best model to describe the spread in the measured variable. Figure A-4 presents a plot of the probability density function (pdf) for the lognormal distribution. Many variables in nature are lognormally distributed (for example the height of people). The log normal distribution is used when the variable of interest has a known physical lower limit of zero. A variable, x, is lognormally distributed if ln(x) is normally distributed. The lognormal distribution is defined using two parameters, $\mu_y$ and $\sigma_y$.

Let the original data be defined by the variable x with mean and standard deviation as:

$\mu_x$ = mean of the measured data for variable x

$\sigma_x$ = standard deviation of the measured data for variable x

Define a new variable y as:

$y = Ln(x)$ (Eq A-10)

Then variable x is said to be lognormally distributed if y is normally distributed. We can now write the mean and standard deviation of y ($\mu_y$ and $\sigma_y$.) [Ref. A-3, A-4 and A-5] as

$$\mu_y = Ln\left[\mu_x^2 /((\sigma_x^2 + \mu_x^2)^{1/2})\right] \quad \text{(Eq A-11)}$$

$$\sigma_y = \left[Ln(\sigma_x^2 / \mu_x^2 + 1)\right]^{1/2} \quad \text{(Eq A-12)}$$

Where:

Ln is the natural logarithm (base e) of the argument

**Appendix A References:**


A-1. P. L. Meyer, *Introductory Probability and Statistical Applications*, 2nd Edition, Addison-Wesley, Reading Mass, Library of Congress Catalog no. 75-104971, 1972, pp. 74

A-2. D. Vose, *Risk Analysis–A Quantitative Guide*, 3rd edition, Wiley, Chichester, West Sussex, England, ISBN 978-0-470-51284-5, 2008, pp. 686-687

A-3. "Lognormal mean and variance - MATLAB lognstat". www.mathworks.com. Retrieved 27 December 2018

A-4. M. Mood, F. A. Graybill, and D. C. Boes. *Introduction to the Theory of Statistics.* 3rd ed., New York: McGraw-Hill, ISBN 0-07-042864-6, 1974. pp. 117.

A-5. E.L. Crow and K. Shimizu, Lognormal Distributions – Theory and Applications, Marcel Dekker, New York, ISBN 0-8247-7803-0, 1988, pp. 9

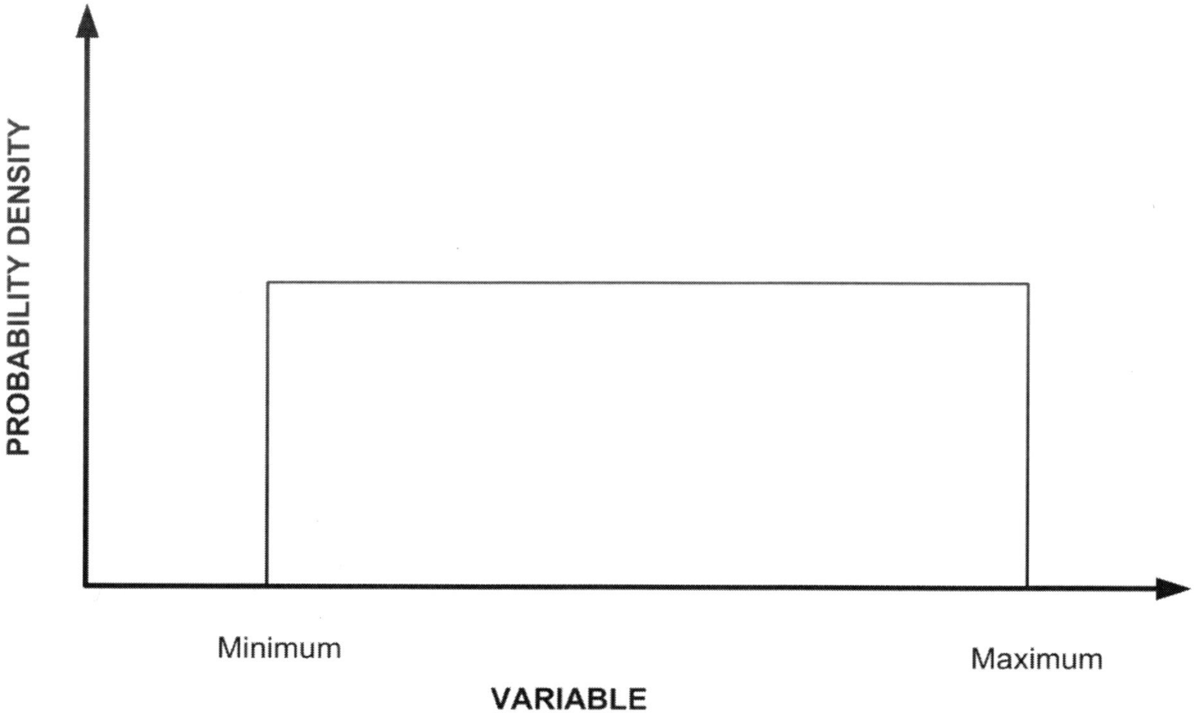


Figure A-1. Uniform Probability Distribution

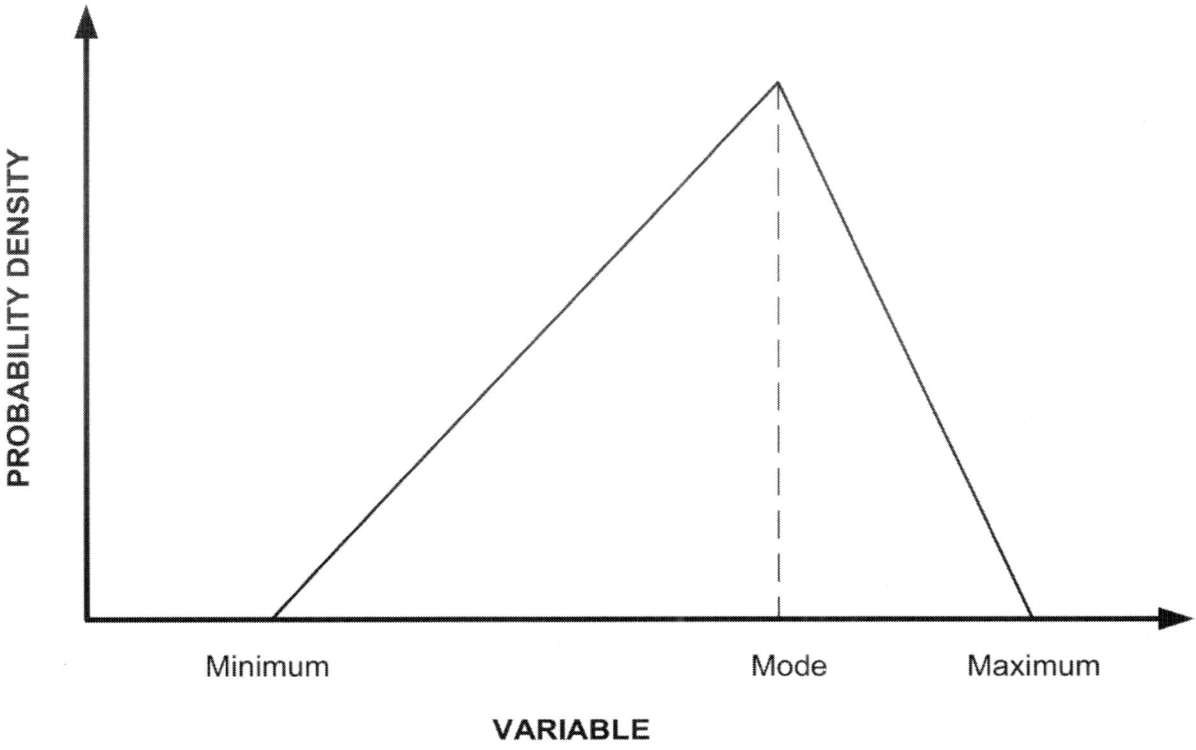


Figure A-2. Triangular Probability Distribution

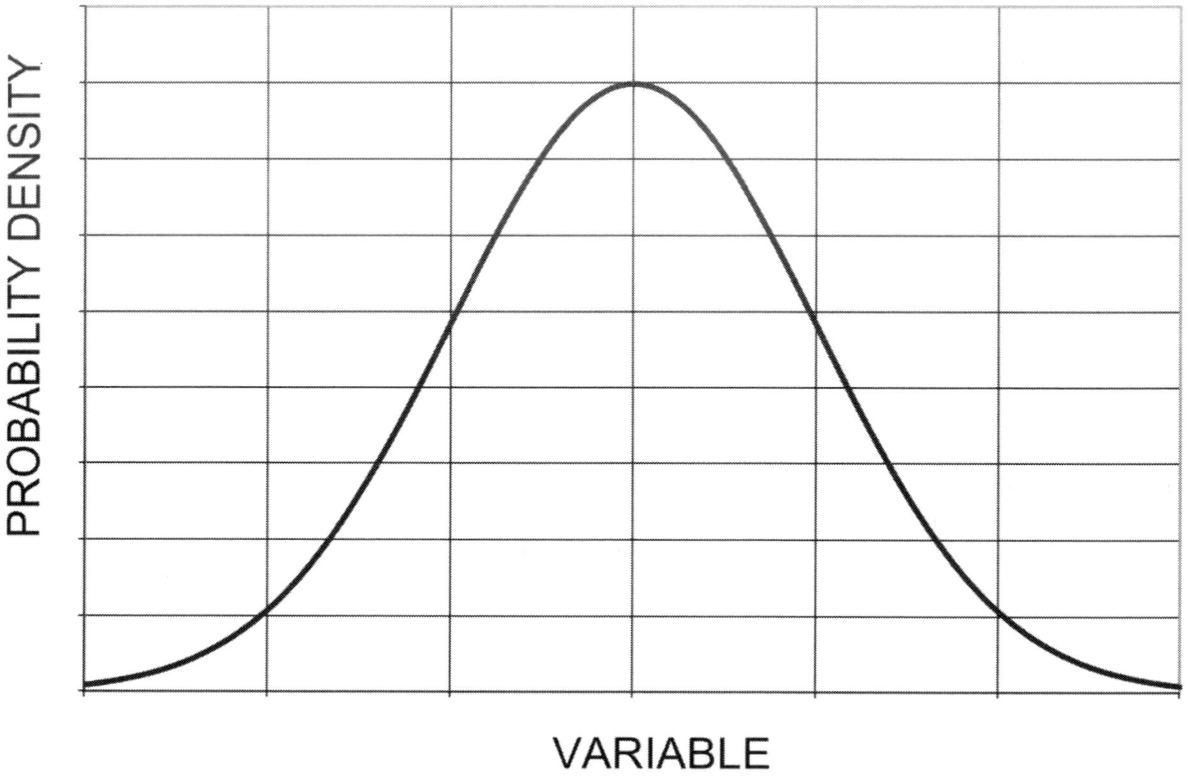


Figure A-3. Normal Probability Distribution

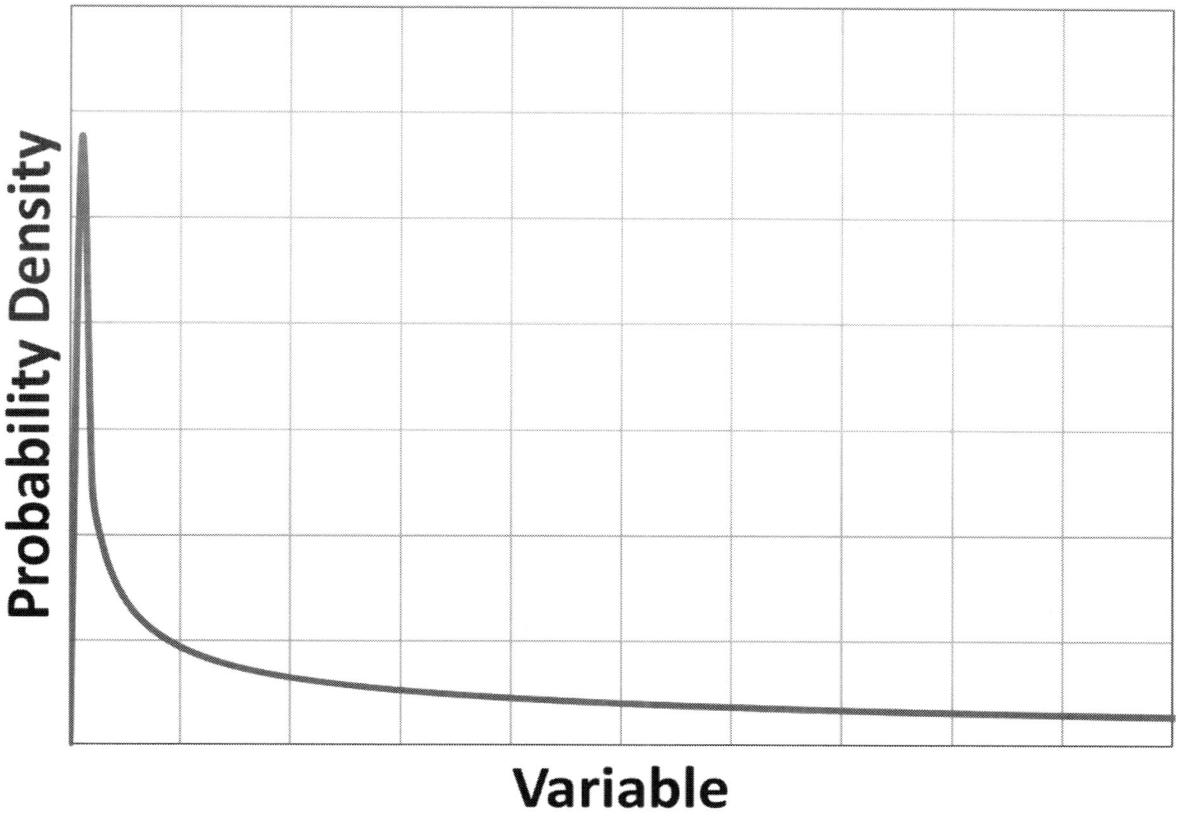


Figure A-4. Lognormal Probability Distribution

# 590424 Making the Most of Your PHA

**Abstract:**

A Process Hazard Analysis (PHA) is required to address OSHA Process Safety Management (PSM) and EPA Risk Management Program (RMP) regulations. Understanding the potential process hazard is widely recognized as a foundational element of PSM and RMP, and in addition to simply meeting regulatory requirements, can be a critical mechanism for flushing out potential vulnerabilities early in the design process for capital projects. Although an important activity, PHA is also recognized as a sizeable investment in time and energy for an Owner/Operator. So why not make the most of it? Other important activities at plants such as Mechanical Integrity, Alarm Rationalization, and Management of Change are able to stay focused and possibly be streamlined (i.e., reduced cost) based on the ability to use the results of a quality PHA having the necessary information and usability of documentation. So why not structure the PHA and documentation to be able to capitalize on potentially significant cost-savings and possible program integration?
The thrust of this paper and conference presentation will be to identify tips for conducting and documenting the PHA to maximize its ability to be used for supporting other important programs such as MOC, MI, and Alarm Rationalization. The objective will be to allow PHA Coordinators to be able to readily apply these tips to bring about improvements and cost-savings on these interrelated programs.

# 589814 Theoretical Analysis of the Viability of Isopropanol-Cyclohexane Solvent Systems for Biomatter Lipid Extraction

**Abstract:**

A theoretical alternative solvent combination for the extraction of lipids from biomatter is the Water, Isopropanol, and Cyclohexane instead of the more common combination used in the Bligh and Dyer Method (Water, Methanol, and Chloroform). While the Bligh and Dyer Method is considered effective, toxicological concerns have been raised over its use of both Chloroform and Methanol. The alternative method utilizes solvents that do not pose as high as a toxicological concern while theoretically maintaining a similar level of practical viability. The alternative method's extraction efficacy was judged against the performance of the Bligh and Dyer method using the simulation software Aspen Plus V10®. The equilibrium data generated from Aspen Plus V10® was then graphed into tertiary diagrams for data analysis using the graphing software SigmaPlot.

# 592166 Techno-Economic Analysis of the Conversion of CH4 and CO2 Intoacetic Acid Using Microwave Plasma

**Abstract:**

Methane and carbon dioxide together represent 90% of greenhouse gases [1], and one way of diminishing their emissions is through the development of pathways that convert them into added value chemicals. Acetic acid is one of such chemicals, and efforts at the laboratory scale have been carried out to obtain alternative ways of producing it. However, there are very few studies that evaluate the possibility of implementing these processes at a larger scale. In this paper, techno-economic and sustainability analyses were used to assess the feasibility of upscaling two pathways to produce acetic acid: (1) by direct conversion of methane and carbon dioxide and (2) by indirect conversion, using syngas (obtained from dry methane reforming) as an intermediate. For the direct method a microwave non-catalytic plasma reactor was considered as the main driver in the process simulation [2]. For the indirect method, a zeolite catalytical single step reaction unit was considered [3]. APEN Plus was used to simulate the reaction and purification systems of both methods, and alongside literature values and plant cost estimation techniques, the capital and operational expenses were calculated. Finally, the feasibility of both projects is assessed on carbon sequestration mass balances and economic metrics. The main sensitivity variables that need to be improved in order to obtain a minimum ROI threshold value were also evaluated.

Key words: carbon management, process design, microwave plasma reactor

[1] EPA. *Overview of Greenhouse Gases.* United States Environmental Protection Agency. 2017.

[2] Wang, L., Yi, Y., Wu, C., Guo, H., Tu, X. *One-Step Reforming of CO2 and CH4 into High-Value Liquid Chemicals and Fuels at Room Temperature by Plasma-Driven Catalysis.* Angewandte Chemie. 2017.

[3] Lopez, L., Velasco, J., Montes, V., Marinas, A., Cabrera, S., Boutonnet, M., Jaras, S. *Synthesis of Ethanol from Syngas over Rh/MCM-41 Catalyst: Effect of Water on Product Selectivity.* Catalysts. 2015.

# 589668 Model Predictive Control Design for the Allam Power Plant Integrated with an Air Separation Unit

**Abstract:**

Since electricity and heat generation account for around 25% of this greenhouse gas emission, high efficiency power generation technology such as the Allam cycle which emits essentially no $CO_2$ and NOx is much needed. Being an oxy-fuel power cycle, the Allam power plant is inherently integrated with an air separation plant (ASU). In this study Aspen Plus Dynamics was used to model the integrated Allam power plant/Air Separation Unit (ASU) with a high degree of heat and work integration based on the steady state work of Net Power. In our previous work, regulatory and advanced PID controls were implemented and tested for various set point tracking and disturbance rejection scenarios. However, PID controllers have limitations for various reasons. Consequently, a model predictive control (MPC) was designed in this study to optimize and improve the performance of the integrated unit and to test its performance in terms of dynamic behavior and robustness to model uncertainties.

# 589672 Explore the Function of Ion-Exchange Membrane (IEM) in Membrane Capacitive Deionization (mCDI) Cell, an Alternative Technique in Brackish Water Desalination Field

**Abstract:**

In the arid west, the freshwater supply of many communities is limited, leading to increased interest in tapping brackish water resources. Although reverse osmosis is the most common technology to upgrade saline waters, there is also interest in developing and improving alternative technologies. Here we focus on Capacitive Deionization (CDI), which has attracted broad attention as an energy efficient desalination technology, especially for ion extraction from low salinity brackish water. The objective of this work is to explore the function of ion-exchange membrane (IEM) in enhancing the performance of CDI cell. In this study, a fully coupled two-dimensional model describing ion transport and adsorption behavior in Membrane Capacitive Deionization (mCDI) cell was developed and solved numerically, taking advantage of Darcy's Law, Nernst-Planck equation, and Donnan equilibrium considering hydraulic dispersion in channel, non-ideal membrane, and modification of ion diffusivity in porous media. The model behavior was supported by the experimentally observed variations in transient effluent concentration under different operating conditions. The results showed that ion flux was significantly improved because of the IEM, leading to a faster desalination rate. The equilibrium adsorption amount is increased due to the permselectivity of the IEM, resulting in extra ion accumulation in electrode's macropores during desalination process. The cut-off time (optimal time to switch polarity to generate highest energy efficiency) of mCDI was shorter with a higher salt removal percentage compared to CDI under the same operating conditions. A sensitivity study of the thickness and the fixed charge density (starting from 500mM) of the IEM was done separately, and neither of them were found to be the dominant influencing factors under a low concentration range of brackish water (below 100mM) in this study.

# 590103 An Integrated Process for the Production of Propylene Glycol from Shale Gas

**Abstract:**

Propylene glycol (PG) is produced on a large scale primarily for the production of environmentally safest and least toxic industrial and domestic products such as polyester resins, thermoset plastics, polyurethanes, food additives and antifreeze. In this research, a novel integrated plant is developed for the production of PG from shale gas. This integrated approach has the benefit of safer operating conditions since the intermediate propylene oxide, which is explosive, does not need to be stored and transported. Furthermore, there are potential economic benefits from integration. The overall plant is simulated in the Aspenplus environment, and a variety of process conditions are tested at steady-state to optimize the production of PG. Heat-integration tools are utilized for energy-saving and capital cost reduction opportunities. A comparative economic assessment based on existing plant information indicates that the use of process integration techniques has the potential to reduce costs significantly.

**Keywords**: Process Integration, Heat Integration, Propylene Glycol

# 592456 Monitoring Methods for Fault Detection and Diagnosis: Algorithms and Applications

**Abstract:**

Many industrial processes collect an abundance of data from different sensors. These sensors often measure a wide variety of physical properties in order to ensure that these parameters are adhering to expected values. This is essential to ensure plant and operator safety, increase economic benefits, and maintain product quality.

A number of algorithms have been developed in order to improve existing fault detection and diagnosis performance. These algorithms integrate a number of different data-driven driven tools and methods. Multiscale wavelet-based representation of data can be used in order to handle data that is autocorrelated, non-Gaussian, and noisy. Hypothesis testing methods such as the Generalized Likelihood Ratio (GLR) technique can be used in order to provide the best possible detection for a fixed false alarm rate. Moreover, certain model-based methods have been developed in order to monitor process drifts and degradations in the process model, even when a process is operating under control.

This work will discuss the different tools utilized in development of the algorithms, their features, and practical applications, including the benchmark Tennessee Eastman Process (TEP).

# 589902 Selective Separation of Industrial Relevant Metal Ions from High Tds Water Using Wafer-Enhanced Electrodeionization (WE-EDI)

**Abstract:**

There is a need for an efficient and economical process of ion-selective separation. In hydraulic fracturing, for instance, several million gallons of water are used in a site, most of which are disposed of by deep well injection taking the water from the ecosystem and causing earthquakes. However, due to the high amount of ions in wells at fracking sites, pressure driven processes for water recovery are impossible to use. This project will demonstrate an emerging environmentally friendly Wafer Enhanced Electrodeionization (WE-EDI) technology which attempts to eliminate the shortcomings that come with normal EDI and separates effluent streams into concentrated ions and clean water. In WE-EDI, ion-exchange resins are replaced by wafers that consist of cation and anion exchange beads and a polymer binding agent. In this study, selective separation of these ion exchange wafers and the effect of electroosmotic drag on the membrane structure have been performed for various anions and cations (Na+, K+, Ca2+, Cl-) and improved selectivity and higher water recoveries have been observed while significant modeling has taken place to predict the selective separation as a strong function of the equilibrium constants between the ions and the beads. This project is supported by the Membrane Science and Technology (MAST) Center and is conducted with the collaboration and guidance of industry mentors.

# 593303 Influence of Particles on Amine Losses during $CO_2$ Capture: A Process Simulation Coupled Aerosol Dynamics Model _

**Abstract:**

**Influence of Particles on Amine Losses during $CO_2$ Capture: A Process Simulation Coupled Aerosol Dynamics Model**

David Dhanraj and Pratim Biswas

Aerosol and Air Quality Research Laboratory

Department of Energy, Environmental and Chemical Engineering

Washington University in St. Louis

St. Louis, MO 63130, USA

Efforts to reduce the concentration of atmospheric $CO_2$ have been initiated due to escalating concerns of global warming and climate change. Amine-based post-combustion $CO_2$ capture offers an economically viable retrofit for power-plants in combating these emissions. However, solvent losses have been a major challenge to commercializing this technology. Recent experiments[1] show that vapor-based and aerosol-based solvent losses are the key mechanisms of amine solvent loss during $CO_2$ capture. While vapor-based amine losses can be abated by employing techniques like water-washing, there has been little work to curtail aerosol-based solvent losses. This is primarily because the underlying mechanisms governing these losses are still not well understood. Although there are a few numerical models proposed for this system, there are no models that consider aerosol phenomena such as coagulation or particle loss. Moreover, these models have not been fully validated through experiments.

Furthermore, there are a few techniques that are being developed to tackle solvent losses. It is known that reducing particle number concentration in the flue gas will reduce amine emissions. Conventional Electrostatic Precipitators (ESP) are however inefficient in capturing particles in the size range 0.1 – 1.0 μm. Amongst different approaches demonstrated, introducing a pre-charging stage and a photo-ionization stage has shown potential in increasing the charging efficiency.

Therefore, the broad objectives of this work are primarily focused on two facets of the amine loss problem. The first is the development of a model that integrates aerosol phenomena with a process engineering model that accounts for the reactive mass transfer process in a packed-bed absorber. The second is the design, fabrication and testing of a bench-scale Photoionization Enhanced Two-staged Electrostatic Precipitator (PI-ESP) at Abbott Power Plant. In essence, with the PI-ESP aimed at reducing the particulate concentration in the flue gas, and the model relating the influence of particulate concentration in the flue gas to amine emissions, this work aims to illuminate the underlying mechanisms governing amine emissions, and demonstrate a solution to tackle the problem.

**References**

1. International Journal of Greenhouse Gas Control, vol. 85, pp. 86-99, 2019.
2. International Journal of Greenhouse Gas Control, vol. 19, pp. 138-144, 2013.

# 588750 Advanced Process Control Continuous Improvement

**Abstract:**

Advanced Process Control Continuous Improvement

Several articles report that after an 18-24 month period over half of the Advanced Process Control (APC) installations are performing at either pre-installation levels or have been removed. There are diverse reasons for the performance shortfall; a major one is that the process characteristics drift from those that generated the controller model. The core of APC is the controller model, typically a linear local representation of the process (gains and dynamics), which is obtained by best matching the model to process responses from incremental changes in the manipulated and disturbance variables. But, when throughput rates and other attributes of the process (reactivity, efficiency, fouling, feed components, etc.) change, the model no longer represents the process, control degrades, and the benefit of the APC diminishes. In this case, recalibration of the model (retesting the process response and adjusting model coefficients to best match) can return the APC to full functionality

A procedure has been developed that employs dimensionless equations to compute the optimum recalibration interval. This optimum recalibration interval is based on a form of decline in performance that is modeled based on the owner-user performance criteria. Further, the dimensionless numbers can be used to compare the performance across units of similar types and across industries. The form of the equation allows for quantitative processes to change the interval for continuous improvement in the recalibration procedure. This continuous improvement process is based on learning curve theory.

This presentation extends that evaluation to the continuous improvement process. Specifically, this presentation looks at four aspects of the continuous improvement cycle. First the for the cost of recalibration is evaluated in light of experience curves. Specifically, a model for evaluating the expected progress of recalibration costs. Second, the income is evaluated for learning. For every recalibration to bring the APC back to functionality, there is learning that takes place. A proposed model is presented for this progression of learning Third, an overall yield equation is developed for an APC. This yield equation contains three aspects. The pre-APC variance, the reduction in the standard deviation for implementing the APC and the standard error on which limits are based. Fourth this overall yield equation is used to show there is a limit to the overall yield and thus the income for subsequent applications for learning to the income.

# 590374 Missed Opportunities in Process Control

**Abstract:**

There has been an order of magnitude increase in measurement accuracy and controller capability. There is a resulting opportunity to do much more in terms of increasing process performance through modeling and control. Dynamic models can be used to provide a safe and effective means of experimentation leading to innovation. Significant knowledge and resulting advances in process control are the result of using such models which today employ a digital twin to enable use of the configurable controllers and operator interface deployed in the actual plant. Key blocks are developed for modeling instrumentation dynamics. Examples are provided of what has been explored and discovered to get the best control valve response and rangeability and to eliminate self-inflicted disturbances from discontinuities, limit cycles, resonance and interaction. The solutions can be simple to understand and to implement.

# 590624 Flare Control Options Under the New Flare Regulations

**Abstract:**

Regulations requiring improved monitoring and control of flares at petroleum refineries are starting to be passed along to other manufacturing sectors. At refineries, flares are no required to maintain 270 Btu/scf in the combustion zone, which takes into account the amount of assist gas being provided to the flare. These regulations follow multiple U.S. EPA enforcement actions which alleged various refineries and petrochemical plants failed to adequately control their flare. This presentation will look at various flare control techniques which may be used to comply with the new flare operating requirements, as well as the unintended consequences of flare minimization which have contributed to the control challenges now facing industry.

# 583094 Mild Thermal Cracking of Oil Sands Bitumen

**Abstract:**

A significant portion of Canadian bitumen is transported via pipelines by blending the high-viscosity bitumen with diluent at a ratio of 25-50 vol% to meet pipeline specifications for viscosity (350 cSt at 7°C) and density (940 kg/m$^3$ at 15.6°C). In addition to being expensive, diluents occupy valuable pipeline capacity and must be separated in refineries only to be transported back to bitumen production sites for reuse. Raw bitumen is sold at a discount price on global markets and has limited access to current and new markets due to its low quality. A number of recent studies have shown that partial upgrading technologies, especially if integrated with upstream operations, can effectively overcome these challenges to reduce, or even eliminate the use of costly diluent, increase bitumen quality and market value, potentially decrease GHG emissions, and reduce capital and operating costs compared to bitumen full upgrading. Currently, most of the partial upgrading technologies under development are based on thermal processing. However, due to some undesired properties of thermally cracked products, such as incompatibility and high olefin content, commercial visbreaking processes cannot convert bitumen directly into a product that meets pipeline specifications. In this study, we focused on the investigation of product qualities at different thermal cracking severities to achieve maximum reduction of density, viscosity and total acid number (TAN), while ensuring reasonable product compatibility and stability.

# 589587 K-Saat, a Break-through Solid Acid Alkylation Technology

**Abstract:**

Globally, the demand for cleaner burning gasoline blend components is increasing while overall demand for gasoline is declining. The increase in demand for cleaner burning gasoline is driven primarily by environmental mandates and high performance engines. Alkylate, isomerate, ethanol and ethers represent the cleaner burning gasoline blend components. Among these, alkylate is the ideal clean fuel component because it has a high octane rating, exhibits low vapor pressure, low sulfur levels, no aromatics or olefins. Alkylate is currently produced using either sulfuric or hydrofluoric acid as the catalyst. Both acids are dangerous and can lead to corrosion-induced accidents and impede firefighting efforts. Solid acids have for many years promised safer and cleaner alkylation. The poor stability of most solid acids, however, have resulted in expensive processes with complex reactors and large catalyst inventories, making them uncompetitive with liquid acid technologies.

After years of development, a new solid-acid catalyst technology ExSact has reached the point of outperforming these liquid acids. The engineered solid-acid catalyst has been designed to give over 24-hr cycle-times, robust resistance to typical poisons, the ability to handle a variety of feedstocks (ethylene to amylene), a simple adiabatic fixed-bed reactor design and regeneration using hydrogen while increasing he overall octane pool. The ExSact catalyst forms the core of an inherently safe and efficient K-SAAT process that generates high-octane alkylate. K-SAAT reactors are the best option available to refiners worldwide to move away from HF Alkylation, offering a commercially proven process with high return on investment and vastly diminishing the risk profile. This is achieved by replacing/debottlenecking existing liquid acid alkylation reactors and reusing the existing recovery system (DIB/Debutanizer/Depropanizer/Isostripper).

The stable catalyst performance greatly simplifies the overall process design, which reduces the capital cost of the alkylation plant, while lowering energy consumption. KBR has licensed K-SAAT units and will discuss overall commercialization status in addition to performance data from the operating unit.

# 589635 A Practical Application of Detection-Based Multiclass Classification of Faults: Tennessee Eastman Process

**Abstract:**

## A practical application of detection-based multiclass classification of faults: Tennessee Eastman Process

*Nour Basha, M. Ziyan Sheriff, Costas Kravaris, Hazem Nounou, and Mohamed Nounou*

**Abstract**

The swift detection and accurate classification of faults is crucial for time-sensitive industrial processes in order to meet minimum standards of quality and production and prevent irreversible damage to valuable equipment. The problem of multiclass classification is of great interest to literature nowadays, due to its heavy reliance on advanced concepts of machine learning and data mining.

Model-free process monitoring techniques, such as Principal Component Analysis (PCA), have proven to be an excellent resource to researchers for applications involving the detection of faults [1,2,3]. That is due to their ability to filter out noise, extract correlations between variables and detect even small deviations from normal operating conditions. In addition, sophisticated statistical methods, such as the Generalized Likelihood Ratio (GLR), have frequently been used in tandem with process monitoring techniques in order to significantly improve their fault detection rate [4,5].

This work extends the application of a combination of efficient process monitoring techniques towards the problem of multiclass classification, namely the classification of faults found in the benchmark Tennessee Eastman (TE) process [6,7]. Our proposed method also makes use of established machine learning methodologies in order to map an array of fault detection based decisions to a single classification decision per sample. The precision of our proposed method is compared to the performances of multiple deep learning techniques on the same dataset [8] in order to demonstrate a more tangible argument of its efficacy and validate its performance.

**References**

[1] I. T. Jolliffe, Principal Component Analysis, Springer, 2002.

[2] P. Sanguansat (Ed.), Principal Component Analysis: Engineering Applications, InTech, 2012.

[3] P. Sanguansat (Ed.), Principal Component Analysis: Multidisciplinary Applications, InTech, 2012.

[4] S. Wang, M. R. Reynolds-Jr., A GLR control chart for monitoring the mean vector of a multivariate normal process,

*Journal of Quality Technology* 45 (2013) 18–33.

[5] M. Z. Sheriff, M. N. Karim, H. Nounou, M. Nounou, Process monitoring using PCA-based GLR methods: A comparative study, *Journal of Computational Science* 27 (2018) 227–246.

[6] J. J. Downs, E. F. Vogel, A plant-wide industrial process control problem, *Computers & Chemical Engineering* 17 (1993) 245–255.

[7] C. A. Rieth, B. D. Amsel, R. Tran, M. B. Cook, Additional Tennessee Eastman Process Simulation Data for Anomaly Detection Evaluation (2017). doi:10.7910/DVN/6C3JR1.

[8] S. Heo, J. H. Lee, Fault detection and classification using artificial neural networks, *10th IFAC Symposium on Advanced Control of Chemical Processes* 51 (18) (2018) 470–475.

# 589698 Practical Machine Learning for Process Control, Monitoring and Diagnostics

**Abstract:**

Process monitoring and control has historically had a rich influx of data centric approaches. The advent of Big Data and Machine Learning/AI has promised much to enhance current capabilities. In many cases, these approaches have struggled to take off because (a) they are too complex and impractical to maintain in industry and/or (b) they have failed to include the most important ingredient during design and execution – the (domain) subject matter expert (SME) who understands the process. Utilizing SME knowledge and feedback is critical for the success of any advanced analytics framework in industry.

In this talk, requirements for a practical, industrially implementable, machine learning framework are discussed. Three examples are discussed. The first application highlights the capability of a practical machine learning framework to simplify APC maintenance and troubleshooting when used in conjunction with domain expertise. Such an approach can also be leveraged during controller and inferential modeling to identify important features and embed known, domain specific relationship constraints. The second example demonstrates how machine learning methods can be used to simplify APC/RTO LP/QP tuning utilizing economic objectives and operational preferences. The third example highlights how practical machine learning can play an important part in gathering insights from vast volumes of data. Data used to gather insights can be from diverse sources – process data, alarms, events are some common examples.

# 589404 Fired Heater Convection Section Steam Generation Tube Failure (*Understanding the impact of flow regime and high peak heat flux profiles)*

**Abstract:**

Two catalytic reformer unit heaters were facing convection section steam generation tube failures recurring every year. This problem was determined by low-pressure drop across steam generation banks and subsequent IR scanning of tubes. It demanded frequent removal of failed tubes. Initially, caustic gouging, poor tube side flow regime, high localized heat fluxes and improper flue gas flow distribution were thought as a possible reason for tube failure. However, water chemistry was being looked at and was not a contributing factor for the tube failures as other heater receiving the same boiler feed water was not facing tube failure issues.

Further, CFD modeling was adopted to study the flue gas flow distribution between the heaters. A damper was installed in the flue gas duct of the heater to control the flue gas flow distribution. Thus, high peak heat flux and poor tube side flow regimes were identified as the major factors causing tube failures.

- The heaters were operating with circulation ratio in the range 7 to 9 and the peak heat flux in the tube failure region was also very high (25,000 – 50,000 Btu/hr-ft$^2$).
- An unstable slug flow regime was predominating in the tube failure region of both the heaters.

This paper describes the scheme to overcome these issues and eradicate tube failures.

# 589927 A Simulation Workflow for Flow, Thermal, and Vibration Analysis of Hvu Transfer Line

**Abstract:**

**In High Vacuum Unit (HVU), vacuum gas oil is recovered from long residue feed. Vacuum gas oil yield, reliability and operability depend on correct design of the vacuum transfer line. Correct and safe design of Transfer Line also prevent excessive cracking and enable a longer run length before decoking. A simulation model is developed for analysis of HVU transfer line flow velocity and pressure drop profile considering flashing along the line. Cases of both normal operation and start-up conditions are considered. Subsequently, a thermal and a vibration analysis model were also developed to look into the thermal stress and vibration safety of the design. In this presentation, we will discuss the complete workflow of simulation analysis highlighting the critical aspects in the correct and safe design of HVU Transfer Line.**

# 615206 Sponsored Technology Workshop - Capital Asset Lifecycle Management - Siemens

**Abstract:**

Today's Energy & Utilities industry must be more efficient and cost-competitive to succeed. Implementing a lifecycle approach to consolidating project technical information is critical to meeting these challenges. Our solution improves how you manage change, execute work processes, drive quality, and connect to external suppliers and customers.

# 589292 A Practical Application of Model-Based Techniques to Track Equipment Degradation and Process Drifts: Monitoring of Fouling in Heat Exchangers

**Abstract:**

Process monitoring and control is a vital component of many industrial processes, required in order to ensure safety and product quality [1]. Process control allows a process engineer to ensure that different streams adhere to strict production standards, through controllers that ensure set-points are being maintained, despite fluctuations in input values or disturbances.

Unfortunately, it is possible for a given process model to also experience degradation over time, e.g., fouling in a heat exchanger. For a heat exchanger model the overall heat transfer coefficient can provide a good indication of fouling, as it decreases with an increase in fouling. In the presence of fouling, the controller increases the amount of cooling utility required in order to meet the desired set-point. In 1982, a conservation estimate of the cost of fouling in the US due to increased operational costs was predicted to be $180 million [2]. This cost was predicted to be closer to $18 billion in 2019 [3]. Since the process is operating under control, monitoring just the output variables, i.e., outlet temperatures of the hot and cold streams, is insufficient, as the controller will continuously adjust the cooling utility to ensure that the set-point is being achieved despite increased operating costs. Therefore, an efficient algorithm needs to be developed in order to monitor both states variables, and the parameter, i.e, overall heat transfer coefficient. This will enable the detection of any drift in the process model, along with the assessment of its impact. The impact is application dependent and can include economic, safety, and environmental losses.

Previous work proposed the use of model-based techniques, namely two Kalman filters: Extended Kalman Filter (EKF), and Unscented Kalman Filter (UKF), in order to monitor deviations in process model through estimation of the overall heat transfer coefficient, in addition to the two states [4]. A simulated synthetic model of a heat exchanger was implemented in Simulink, in order to demonstrate the application of the proposed algorithm. This included the development of a cost contour profile to monitor the economic impact of fouling to the process. This work extends the proposed algorithm to real data, obtained from a heat exchanger at Texas A&M University at Qatar, in order to demonstrate its validity and further highlight its practical applicability.

**References**

[1] F. Harrou, M. N. Nounou, H. N. Nounou, and M. Madakyaru, "Statistical fault detection using PCA-based GLR hypothesis testing," *J. Loss Prev. Process Ind.*, vol. 26, no. 1, pp. 129–139, 2013.

[2] W. J. Rebello, S. L. Richlen, and F. Childs, "The cost of heat exchanger fouling in the US industries," 1988.

[3] "Economics of fouling," *Advanced heat transfer technologies*, 2019. [Online]. Available: https://fbhx-usa.com/economics-of-fouling. [Accessed: 10-Oct-2019].

[4] M. Z. Sheriff, H. Nounou, M. Nounou, and M. N. Karim, "Monitoring process degradation through operating regime based process monitoring," in *AIChE Spring Meeting and Global Congress on Process Safety: Process Control Monitoring and Analytics*, 2019.

## AUTHOR INDEX